Written primarily for a non-specialist audience, these essays describe contributions made by some of the University of Cambridge's most colourful and able characters in a number of academic disciplines. The essays reveal particularly fertile periods of development and chart voyages of discovery which have occurred all over Cambridge, under group or individual leadership. Approaches vary, from the presentation of historically significant discoveries to the explanation of current research – 'contributions' in the making. The interweaving of the academic lives of Cambridge figures has done much to enrich understanding within and between disciplines, and to influence their development in particular ways. The enthusiasm with which these figures and disciplines are presented will ensure that readers continue their own investigations into the contributions and contributors mentioned here.

Also of interest

CAMBRIDGE MINDS

Edited by

RICHARD MASON

An intellectual portrait of what has been done in a wide variety of fields in the University of Cambridge, including philosophy, economics, anthropology, and the study of English.

'What emerges from these essays as the university's salvation is the very quality which so infuriates tidy-minded bureaucrats and central education planners: that it is not a single, efficiently-planned organisation, but a sprawling, benignly anarchic coalition of three distinct entities . . . When one element fails, there are others to keep talent within Cambridge's capacious fold.'   *The Times*

' . . . a series of essays on the intellectual achievements of Cambridge in the last century or so. They began as lectures at a summer school and are addressed to a "non-specialist audience": but they are all by experts, and the general quality is very high.'   *The Sunday Telegraph*

0 521 45625 8 paperback

A CONCISE HISTORY OF THE UNIVERSITY OF CAMBRIDGE

ELISABETH LEEDHAM-GREEN

'With eight centuries to cover, Elisabeth Leedham-Green, who aims to produce "a standard introduction for anyone interested in" the University of Cambridge, assumes a daunting task. She has succeeded admirably, for this handsomely-produced, well-indexed and well-illustrated "portable" history shows clearly and invitingly how Cambridge evolved from the thirteenth-century seccession of scholars from Oxford into the world-class university we now know.'   *The Times Literary Supplement*

' . . . packed with information, concise but not congested. It carries the reader on, from medieval times to the post-war present; it can and should be read as a whole.'   *Cambridge*

0 521 43370 3 hardback
0 521 439787 paperback

CAMBRIDGE

CONTRIBUTIONS

Edited by
SARAH J. ORMROD

CAMBRIDGE
UNIVERSITY PRESS

University Printing House, Cambridge CB2 8BS, United Kingdom

Cambridge University Press is part of the University of Cambridge.

It furthers the University's mission by disseminating knowledge in the pursuit of education, learning and research at the highest international levels of excellence.

www.cambridge.org
Information on this title: www.cambridge.org/9780521592437

© Cambridge University Press 1998

First published 1998

A catalogue record for this publication is available from the British Library

ISBN 978-0-521-59243-7 Hardback
ISBN 978-0-521-59738-8 Paperback

Contents

Contents

Contributors and contributions

Cambridge has played a central role in the development of astronomy and cosmology over the past fifty years: steady-state cosmology; evidence from radio astronomy that there was a big bang; the discovery of pulsars and current research on black holes and very distant parts of the universe are all part of the University's contribution. MARTIN REES is Royal Society Research Professor at the Institute of Astronomy and, since 1995, Astronomer Royal. He held the Plumian Chair of Astronomy and Experimental Philosophy from 1973 to 1991 and was also Director of the Institute of Astronomy for nine of those years. Professor Rees is a member of many international learned societies and has published a range of articles and reviews in scientific journals. His popular book, *Before the Beginning: Our Universe and Others*, was published in 1997.

The University of Cambridge has made significant contributions to medicine since the 1540s, when a Chair of 'Physic' was established by Henry VIII. The contributions of some of the many figures in this field, such as William Harvey, the sixteenth-century discoverer of the circulation of the blood, and Frederick Gowland Hopkins who discovered vitamins in the 1920s, are assessed by MARK WEATHERALL. Dr Weatherall is co-author of *Dynamic Science: Biochemistry in Cambridge, 1898–1949*.

Although Cambridge can number amongst its alumni an impressive roll-call of famous writers, including Spenser, Marlowe, Milton, Wordsworth, Tennyson and Byron, there was no degree in English at the University until 1917. STEFAN COLLINI discusses the rapid rise in the influence of 'Cambridge English' thereafter, through figures such as I. A. Richards and F. R. Leavis. Dr Collini is Reader in Intellectual History and English Literature at Cambridge. His recent publications include *Matthew Arnold: A Critical Portrait, Public Moralists: Political Thought and Intellectual Life in Britain, 1850–1930*, and *That Noble Science of Politics: A Study in Nineteenth-Century Intellectual History.*

GEOFFREY HARCOURT is Reader in the History of Economic Theory at Cambridge and a Fellow of Jesus College. In 1988, he was made Professor Emeritus of the University of Adelaide, and in 1994 was elected as an Officer in the General Division of the Order of Australia (AO) for services to economic theory and to the history of economic thought. He was made the 1996 Distinguished Fellow of the Economic Society of Australia. Dr Harcourt has published and edited a vast range of essays and books on economics over the past forty years, most recently producing *A 'Second Edition' of The General Theory* (edited in two volumes with P. A. Riach), *Capitalism, Socialism and Post-Keynesianism, On Political Economists and Modern Political Economy*, and *Post-Keynesian Essays in Biography*. His chapter considers the historical development of economics at Cambridge, and the work of John Maynard Keynes and his followers.

GILLIAN SUTHERLAND is Fellow, Gwatkin Lecturer and Director of Studies in History at Newnham College, one of Cambridge's three women-only colleges, whose foundation at the end of the nineteenth century marks an early chapter in the story of the arrival and acceptance of women at Cambridge. Even though women were not admitted as full members of the University until 1948, very considerable contributions were made by women at Cambridge before this milestone was reached. Dr Sutherland explores the nature of these from the inter-war years. She is currently working on a double biography of A. J. and B. A. Clough.

There have been major contributions by Cambridge scholars to the study of Classics for well over three hundred years. Textual critics Richard Bentley, Richard Porson and A. E. Housman stand out as giants in the field at Cambridge in the eighteenth, nineteenth and twentieth centuries, and their influences survive today. PAUL CARTLEDGE focuses in his chapter on the future of classical study at Cambridge. Dr Cartledge is Reader in Greek History and a Fellow of Clare College, Cambridge. Amongst his recent publications are *Aristophanes and his Theatre of the Absurd*, *The Greeks: A Portrait of Self and Others* and *The Cambridge Illustrated History of Ancient Greece*.

PETER LIPTON is Head of the Department and Professor of History and Philosophy of Science, and a Fellow of King's College, Cambridge. Having completed his doctorate at Oxford, Professor Lipton taught at Clark University and Williams College in Massachusetts, before coming to Cambridge. His interest is in general philosophical questions about how science works and what it achieves. His chapter addresses the development of the history and philosophy of science in Cambridge and discusses a philosophical puzzle about the way scientists test their theories.

MADELEINE ARNOT is a University Lecturer in Sociology in the Faculty of Education, and a Fellow of Jesus College, Cambridge. Dr Arnot was Noted Scholar at the University of British Columbia, Vancouver and a visiting scholar at the University of Tromsø, Norway. She has directed research projects for the Office of Standards in Education and the Equal Opportunities Commission and has published extensively on educational equality. Dr Arnot was awarded a Leverhulme Research Fellowship to extend a European project on gender and citizenship. Her chapter draws on the findings of this research and that of her colleagues to illustrate the challenges which the European Union provides for educational researchers and practitioners.

JOHN PARKER is Director of the University Botanic Garden and Professor of Plant Cytogenetics, based in the Department of Plant Sciences in the University of Cambridge. He is also a Trustee of the Royal Botanic Gardens, Kew and a Research Fellow of the Natural History Museum, London. He was previously at the Universities of Oxford, London and Reading before coming to Cambridge in 1996. Professor Parker carries out research on the structure, organisation and behaviour of chromosomes in both wild populations of plants and crop species. His contribution to this collection of essays considers the origin of the University Botanic Garden, its current activities and the roles the Botanic Garden will play in the next century.

As the title of his chapter suggests, HERBERT HUPPERT discusses some of the major figures, past and present, in the field of Cambridge geophysics as well as the contributions they made to our understanding of the structure and evolution of the Earth. Professor Huppert is Professor of Theoretical Geophysics and Foundation Director of the Institute of Theoretical Geophysics. He is also a Fellow of King's College, Cambridge and of the Royal Society. Professor Huppert has been a visiting research scientist at the Australian National University, University of California, San Diego, Massachusetts Institute of Technology and the University of New South Wales. His many published papers are on topics as wide-ranging as applied mathematics, fluid mechanics, geology, geophysics, oceanography and meteorology.

One of Cambridge's less reputable contributions to twentieth-century history is as a training ground for the most gifted of communist spies in the 1930s: Anthony Blunt, Donald Maclean, Guy Burgess and Kim Philby. A leading authority on espionage in this period, CHRISTOPHER ANDREW, is Professor of Modern and Contemporary History in Cambridge and a Fellow of Corpus

Christi College. His *Secret Service: The Making of the British Intelli-gence Community* is a standard work. With Oleg Gordievsky he has written *KGB: The Inside Story of its Foreign Operations from Lenin to Gorbachev*. More recently he has produced *For The President's Eyes Only: Secret Intelligence and the American Presidency from Washington to Bush*.

Preface

The chapters in this book are based on lectures given by their authors as part of the University of Cambridge International Summer Schools in July 1996 and July 1997. Some 300 people of all ages and from over thirty countries take part in these inter-disciplinary programmes each year, whilst a further 850 attend a range of concurrent subject-specialist Summer Schools. Those attending the International Summer School spend a month in Cambridge, living in the colleges and gathering in smaller groups for classes three times a day. In addition to classroom sessions and private study, the entire group is invited to attend a plenary lecture each day.

In my capacity as Director of the Summer School, I arrange a series of plenary lectures which will stimulate this audience for an hour each morning and encourage its members to think about things beyond their immediate course of study. I am struck each year by the quality of these lectures, and regret that more people cannot benefit from their preparation. For the past few years these plenary lectures have been open to members of the university, and selected lectures from the 1993 series were published as *Cambridge Minds*. This volume follows both that example and, to a certain extent, a much earlier one: the lectures on astronomy which James Stuart delivered to the workmen of Crewe in the summer of 1868 were written up by him and published by the University Press at the end of 1869, so that a permanent record might be kept and enjoyed.

In selecting the theme for the 1996 and 1997 series I chose not to focus exclusively on single individuals, but rather to address particular disciplines, and asked speakers to elaborate on the contributions made by a number of individuals based at the University of Cambridge to those fields of study. Hence, *Cambridge Contributions*. The plural is intentional since it would be outrageous to suggest that one volume of this type could set itself up to define *The* Cambridge Contribution. It is instead a selection, a sampling, of contributions made by this institution and a few of its people to a selection of disciplines. Not all of the subjects featuring in the Summer School series could be represented here; there were also, for example, excellent lectures on mathematics, molecular biology, philosophy, Quaternary research, social anthropology, computer engineering, history, religious studies and criminology. It would take many more lecture series and many more volumes of this kind to achieve a proper balance of subjects represented, and only then might the cumulative offering gradually edge just a little closer to reflecting *The* Cambridge Contribution.

However, the papers collected in this volume represent a fascinating introduction to the significance of Cambridge people in a number of different fields. The papers present a variety of approaches, entirely suited to the diverse nature of their audience and attempts have been made to preserve this variety as much as possible. The lecturers whose papers appear here were asked to explain the significance of contributions in their own very specialised fields, to a non-specialised audience. Each was encouraged to approach the subject in his or her own way and some chose to talk about their own research, giving the audience the opportunity to witness contemporary 'contributions' in the making. Perhaps the most obvious example of the latter is Madeleine Arnot's chapter, which concentrates on research into European citizenship and

education being undertaken in the University's School of Education – the most 'modern' of all the subject areas addressed in the
collection. Other writers look back on the history of their subjects,
or at more recent developments, and one, Paul Cartledge, bravely
looks forward at the end of his chapter, even beyond the debates
current in the mid 1990s. Stefan Collini's chapter addresses and
questions the reader directly, inviting a reaction in the same way
that the teachers of English at Cambridge asked for a critical
response from their students, as he explains not only the evolution
of the study of English, but also several other ways in which it
might have evolved instead.

The collection has a strong science content. This was a
deliberate attempt, on my part, to broaden the offering given to
our International Summer School students, which in 1996 and
1997 still retained strong leanings towards the liberal arts – a
legacy of the programme's 74-year history. This, in a University
which is known world-wide for its contribution to the sciences,
clearly needed attention and so the history of scientific ideas,
scientific illustration, history of medicine, history of mathematics, environmental studies, philosophy of science and current
perspectives on genetic engineering have now been added to our
Summer School course lists. So whilst several of the chapters
here represent subjects currently taught as part of the Summer
School, the fields of cosmology, Classics and geophysics have
not featured on the curriculum as yet. Nor, in this particular
Summer School, has espionage!

Although the number of women contributors to this collection
has doubled in comparison to that in the 1993 volume (now two,
rather than one!), it could be seen as a reminder, as Gillian Sutherland wrote in 1993, that women 'are still contending for and trying
to act out that "frank recognition"' of likeness between the sexes
which Emily Davies had called upon men at Cambridge to
recognise in 1866. But as Cambridge celebrates the fiftieth anniversary of the admission of women to degrees, many prominent
Cambridge women invited to participate in our Summer Schools

now have exactly the same reasons (international travel on academic grounds, research, vital publication deadlines) for being unable to participate as their male counterparts!

Gillian Sutherland's chapter in this collection concentrates on the difficulties encountered by Cambridge women between the two world wars, and, in spite of these, their considerable achievements. There are references to the contribution of women in other chapters too – Paul Cartledge writes, for example, of Jane Ellen Harrison's contribution to Classical scholarship and in his chapter on medical science Mark Weatherall draws our attention to the Dunn Institute of Biochemistry, which in the 1920s and 1930s was 'one of the few places in Cambridge where your gender didn't matter'.

In this context it seems appropriate to mention the *contribution* Cambridge has made to other institutions in Britain and beyond, either by exporting or, more accurately, exiling several of its brightest women scholars. Gillian Sutherland refers to just two of many: Cecilia Payne-Gaposchkin left Cambridge to be the first Harvard/Radcliffe doctoral student of astronomy, following this with a distinguished career, and Eileen Power, eminent medieval historian, so tired of waiting for due recognition in Cambridge, migrated to the London School of Economics, to receive it there.

It is possible to conjure, from the histories related here, a remarkable assortment of different Cambridge figures. The authors have not simply charted names, dates and achievements, but often given evocative thumbnail sketches which change our perceptions and leave us wanting to learn more. Our imaginations may be caught by the immodest Keynes; Teddy Bullard's pride in the photograph of himself which hung on Herbert Huppert's office door; Malthus's one-liners; Marshall as 'not, in many ways, an admirable man, but a great economist'; Francis Bacon as a politician admitting to accepting bribes, without being influenced by them, and his fatal stuffing of a chicken with snow; the recollections of Porson, a brilliant Classicist, drunk and decidedly disorderly, but still reciting Greek and Latin; the (sober, orderly and) wonderfully apposite orations of James Diggle, continuing in a role established at Cambridge in the 1520s, but proving that Latin can be adapted to express ideas about computer memory stores. Many of the characters are larger than life: Herbert Huppert's description of 'extinct and active volcanoes' applies far beyond the discipline of geophysics.

The chapters often reveal surprising contributions made by figures considered specialists in one field to a different discipline entirely. J. G. Frazer, the anthropologist, A. E. Housman, the poet, and Isaac Newton, the mathematician, all made contributions to Classics. The Classical scholar John Caius used his knowledge of Greek and Latin to translate classical medical texts. We learn of the influence on his own contribution to Cambridge English made by I. A. Richards's earlier training in philosophy, psychology and psychoanalysis. The economist John Maynard Keynes's first book on probability was influential in the field of philosophy of science. And so on. Evidence of the combined creative effect of these many disciplines existing side by side appears time and again.

Just as several of the figures mentioned here achieved prominence in more than one field, there are many instances of contributions overlapping between disciplines. There is the Cambridge tradition of history and philosophy of science, which, as Peter Lipton explains, most obviously embraces what many see as two distinct worlds. And the University Botanic Garden, we are reminded, bridges a number of 'boundaries' between disciplines, providing research resources for zoology, genetics, biochemistry, ecology, to name but a few.

Characters and disciplines interweave. More than once we find characters introduced in a supporting role in one chapter (or in the previous volume) reappearing to take centre stage briefly in another: William Harvey appears briefly in Peter Lipton's chapter, and more prominently in Mark Weatherall's. Newton appears in Martin Rees's piece and Peter Lipton's. Keynes, unsurprisingly, figures largely in Geoffrey Harcourt's chapter on economics, but also has a significant role in Peter Lipton's chapter on the philosophy of science. Anthony Hewish, who wrote for the *Cambridge Minds* volume, is himself discussed as a contributor in Martin Rees's paper. The Revd John Stevens Henslow, at one time holding concurrent Professorships in Mineralogy and Botany, and also retaining a strong interest in natural history, was shown to be enormously influential in the life of the young Charles Darwin, in Richard Keynes's chapter on the Darwins in Cambridge in *Cambridge Minds*. In John Parker's chapter on the Botanic Garden we learn of Henslow's role in the change of direction of the garden's collection: from physic garden to teaching garden. The interweaving of the academic lives of Cambridge figures has done much to enrich understanding across a whole range of disciplines, and to influence their development in particular ways.

In a remarkably concise way these chapters tell much about the evolution of whole Faculties and Departments, or particular ways of thinking or voyages of discovery which have happened all over Cambridge, at particular times, and under particular group or individual leadership. Awareness emerges of particularly fertile periods of change and development in, for example, botany under Henslow in the early nineteenth century; in astronomy under Hoyle and Ryle, Hewish and Bell; in archaeology under Dorothy Garrod and Gertrude Caton-Thompson; in Classics under the Cambridge Ritualists, Cornford, Frazer and Harrison; or in economics under Marshall and Pigou, or Keynes and his 'Cambridge Circus'. Medical science in its broadest sense appears to have flourished at Cambridge in the mid-sixteenth and the late-eighteenth centuries, but Mark Weatherall details specific progress in, for example, physiology, biochemistry and pathology this century. So often these departments, during such periods of recognition, have at their helm pairs of individuals or 'academic mini-dynasties' whose collected contribution is greater than the sum of individual parts (immense though many of those individual contributions may have been). The heady combinations of characters such as I. A. Richards and F. R. Leavis, or of Gerald Lenox-

Conyngham, P. M. S. Blackett, Ernest Rutherford, Harold Jeffreys and Teddy Bullard are just two such examples. There are many others.

Yet Stefan Collini also usefully reminds us that 'luck and timing' play a role in bringing recognition to certain departments for making important contributions to particular disciplines. And this appears to hold especially true of the establishment of the new English course during the First World War, or the vast range of opportunities made available to Cambridge women during the Second World War, or any one of a number of academic friendships which resulted in the hastening of discoveries or the implementation of change.

As organiser of both series from which these chapters have been drawn, I have enjoyed two privileges. The first was to attend every single one of the plenary lectures each year, and to find myself drawn into every new subject, intrigued by the people, their lives and their achievements. The chapters in this volume are – I hope – a representative selection, and a far more precise and lasting memento than the memories of the lectures. I am very much aware that the second privilege, that of acting as editor, has offered a rare opportunity to work more closely with the scholars whose papers appear here. In the best tradition of gifted and immensely busy people, the authors have found time to submit their accounts, interpretations and recollections, making their own 'contributions' in the process. Each has responded graciously and patiently as my cajolings and demands have intruded on lives already full of academic and much larger (and often longer-standing) publishing commitments. And after meeting initial deadlines, those from the first series in 1996 have waited patiently until papers from the second in 1997 were submitted. Each chapter reflects the author's enthusiasm for his or her subject, and for the figures who developed those disciplines. I, and a much larger public than the participants in our Summer Schools, now have access to a permanent record of these collected enthusiasms.

Cambridge Contributions will hopefully be accepted for what it is:

a kaleidoscopic glimpse of just some of the contributions within eleven of the very many different parts of the University's whole, each chapter in itself offering a fascinating range of information and observation. Bringing these papers together has had a similar effect to experiencing the lectures: there are unexpected connections, parallel lives, concurrent dramas. Just as every visitor to an eclectic collection of paintings might be drawn first to a different work, different readers will gravitate at first to different essays. For some, 'thinking European' might be a more relevant and apparently accessible concept than those in the fields of geophysics or astronomy and cosmology. It is not necessary to read the chapters in the order in which they appear here, but I hope that every reader will eventually turn to every chapter, to be rewarded by the discovery that each is immensely readable – witness the ease with which we are guided through the weighing of the Earth in 1775, the calculation of routes of seismic waves, magnetic field observations on the ocean floor to plate tectonics and beyond in Herbert Huppert's chapter on geophysics, or through the life-cycles of stars and the evolution of the universe in Martin Rees's on astronomy and cosmology. The collection offers no more complete a picture than does one room in a very extensive gallery, but within that one room there are exciting and telling resonances between the different contributions. It is hoped that this volume informs and intrigues, and, by giving just a few examples of contributions that Cambridge has made, leads readers to make their own further investigations into the people and work mentioned here.

I am grateful for the support and encouragement of Michael Richardson and my colleagues in allowing me the time to collate this volume, and especially to Maggie Humphries and Shelley Lockwood for all their patience and hard work on the scripts; for the example set by Richard Mason and *Cambridge Minds*; for the enthusiasm of those international students who were the first to enjoy the original lecture series; for the generosity, patience and guidance of the authors themselves; and for advice from William Davies at Cambridge University Press. The lectures on which the chapters were based were supported by the International Division of the University of Cambridge Board of Continuing Education. All profits from the sale of this volume will go to a Scholarship Fund which enables outstanding students from poorer countries to participate in the University's International Summer Schools.

SARAH J. ORMROD
Madingley Hall

CHAPTER I

Some Cambridge contributions to astronomy and cosmology

MARTIN REES

INTRODUCTION

Out of every 10,000 people, 9,999 have something in common: their professional concerns lie on or near the Earth's surface. The 'odd one out' is an astronomer. But a poll taken on King's Parade would yield a higher-than-average chance of finding astronomers: the proportion would still be tiny, maybe one in five hundred, but this is enough to establish Cambridge as a major centre for research in this subject.

SOME EARLY HISTORY

Cambridge's astronomical tradition dates back, of course, to Newton in the seventeenth century. No subsequent individual in the University (nor, perhaps, anywhere in the world of science) has matched his extraordinary gifts and singular achievements. But the tradition in theoretical and mathematical astronomy has been sustained by a succession of eminent names, through to Eddington, Hoyle, and Hawking in the present century.

Cambridge's contributions to the observational side of the subject date back almost as far. In 1702, the Reverend Thomas Plume, Vicar of Malden, in Essex, left his estate to maintain a studious and learned Professor of Astronomy & Experimental Philosophy, and to buy him and his successors books and instruments, Quadrants, Telescopes &c . . . so as any ingenious Scholars or Gentlemen may resort to him in the proper seasons to be instructed and improved by him in knowledge of Astronomy, the Globes, Navigation, Naturall Philosophy, Dialling & other practical parts of the Mathematicks.

An observatory for the Plumian Professors' use was constructed on the King's Gate of Trinity College. But the eighteenth-century incumbents were less diligent than they should have been, and in 1792 the Trustees noted that the 'observatory and instruments belonging to it were through disuse, neglect and want of repairs so much dilapidated as to be entirely unfit for the purpose intended'; the instruments were removed from Trinity in 1797. A similar observatory in St John's (over the gateway between second and third Courts) fared somewhat better, and survived until 1859.

In 1823, a University Observatory was founded on a site 1.5 miles west of the city centre, along the Madingley Road; the original estimate for its construction was £10,000, but it actually cost almost double. The central dome of this impressive neo-classical building originally contained a telescope. The Observatory was also equipped with 'meridian' instruments for determining the accurate positions of stars and planets. The slits through which these telescopes were directed can still be traced in the external masonry. The building was oriented so that they faced exactly south, along a meridian. One telescope was located exactly to the north of Grantchester village church, so that it pointed exactly along the meridian when viewing an oil lamp on the church spire.

It was George Airy, appointed Plumian Professor in 1828, who brought the Cambridge Observatories into effective operation. He was an energetic, indeed brash, character. Before taking up the Plumian Chair he wrote to the Trustees:

The Professor feels confident that the University will not positively require of him, and cannot reasonably expect from his successors, that they will renounce all other expectations and abandon all other sources of income, to employ themselves in occupations so incessantly laborious for so small a stipend as that now attached to the care of the Observatory.

This pleading secured him a pay rise from £300 to £500 per annum. Airy nevertheless held the post for only eight years, before moving to Greenwich (and, no doubt, further enhancing his emoluments) as Astronomer Royal.

One of Airy's achievements was to secure from the Duke of Northumberland, Chancellor of the University, funds for a 12-inch refractor telescope. This instrument survives to the present day, and is still used by undergraduate members of the University Astronomical Society. However its main claim to fame is an inglorious one – it is the telescope that failed to discover Neptune. John Couch Adams, of St John's had calculated, from anomalies in their orbits, that an eighth planet lay beyond Saturn and Uranus. Airy's successor, James Challis, could in retrospect have been the discoverer of Neptune. He had actually recorded its position three times; but, before he got round to analysing his data, Johannes Galle in Berlin had scooped him.

INTO THE TWENTIETH CENTURY: STARS AND ATOMS

After 1900, the west coast of the United States, with its climatic and financial advantages, became the preferred site for large telescopes. The United Kingdom was therefore outclassed in observational astronomy during the first half of the twentieth century. During that period, Cambridge's astronomical strength was on the theoretical side. Arthur Eddington's insights into the nature of stars established him as the pre-eminent astrophysicist of his generation. He was also the prime promoter of Einstein's theories in the United Kingdom, and the leader of the 1919 eclipse expedition which discovered that light rays were indeed bent by the Sun's gravity, as relativity predicted.

An equally versatile and celebrated contributor to astrophysics was Fred Hoyle (one of Eddington's successors as Plumian Professor). Starting in 1946, he built on Eddington's earlier studies of stellar structure to show how the chemical elements – carbon, oxygen, iron, etc. – could all have been transmuted from hydrogen by nuclear reactions inside hot stars.

The abundances of the different kinds of atoms – the basic building blocks for chemistry, and indeed for life – can be measured on Earth, in the Sun, and elsewhere in the Universe. The proportions display striking regularities: for instance oxygen and carbon are common; gold and uranium are rare. We'd like to understand why the atoms were 'dealt out' in these particular proportions. We could leave it at that – perhaps the creator turned ninety-two different knobs. But Hoyle offered a less *ad hoc* explanation.

Stars, especially those that are heavier than the Sun, have complicated life-cycles. They spend most of their lives fuelled by conversion of hydrogen (the simplest chemical element, whose nucleus is just one proton) into helium. When their central hydrogen has been used up, gravity squeezes them further, and their central temperature rises still higher, until helium atoms can themselves stick together to make the nuclei of heavier atoms – carbon (six protons), oxygen (eight protons), and iron (twenty-six protons). A kind of 'onion skin' structure develops: a layer of carbon surrounds one of oxygen, which in turn surrounds a layer of silicon. The hotter inner layers have been transmuted further up the periodic table and surround a core that is mainly iron.

When their fuel has all been consumed (in other words, when their hot centres are transmuted into iron) big stars face a crisis. A catastrophic infall compresses the stellar core to neutron densities, triggering a colossal explosion – a supernova.

The outer layers of a star, by the time a supernova explosion blows them off, contain the outcome of all the nuclear alchemy that kept the star shining over its entire lifetime. There is a lot of oxygen and carbon in this mixture, plus traces of many other elements.

4

The oldest stars formed about ten billion years ago from material that emerged from the big bang. This primordial material contained only the simplest atoms – no carbon, no oxygen, and no iron. Before our Sun even formed, 4.5 billion years ago, several generations of heavy stars could have been through their entire life cycles. Pristine hydrogen was transmuted, inside stars, into the basic building blocks of life – carbon, oxygen, iron, and the rest.

By combining his knowledge of stellar evolution with the data from nuclear physics, Hoyle and his colleagues explained the relative abundances of the different elements. The calculated 'mix' is gratifyingly close to the proportions now observed in our Solar System. Why are carbon and oxygen atoms so common here on Earth, but gold and uranium so rare? The answer involves the 'ecology' of our entire Galaxy – the fate of ancient stars that exploded in our Milky Way more than five billion years ago, before our Solar System formed. To understand ourselves we must understand the cosmos. We are literally the ashes – the 'nuclear waste' – from long-dead stars.

EVOLUTION IN COSMOLOGY

But where did the original hydrogen come from? To answer this question, we must extend our horizons to the extragalactic realm. Our Milky Way, with its hundred billion stars, is just one galaxy similar to millions of others visible with large telescopes.

The overall *motions* in our Universe are simple too. Distant galaxies recede from us with a speed proportional to their distance. Were the galaxies more closely packed together in the past? And do remote galaxies look different, as we'd expect if they were, on average, younger when they emitted the light now reaching us? These questions are important because the answer need not be 'yes' – an expanding universe need not necessarily evolve. This point was forcefully made by Fred Hoyle, together with Hermann Bondi and Thomas Gold, two theorists who had come to Cambridge as refugees from Austria. Bondi, primarily an applied

5

mathematician, contributed influential ideas to astronomy and the theory of relativity. Gold's range of expertise was more eclectic. His academic career was launched by a thesis on hearing and the physiology of the inner ear; he went on to deploy his physical insights in many areas (including neutron stars, as mentioned later).

Bondi, Gold, and Hoyle conjectured that we might live in a 'steady-state' universe, in which continuous creation of new matter and new galaxies maintained an unchanging cosmic scene despite the overall expansion. Individual galaxies would still evolve; but as they aged they would disperse more widely, and new younger ones would form in the gaps that opened up between them. The Universe, having had an infinite past, might have achieved some unique self-sustaining state. The required creation rate was so low that it would have been entirely unde-tectable – one atom per century in each cubic kilometre – but many found the concept *ad hoc* and implausible. Hoyle countered this objection by developing a specific theory to describe how new atoms could occasionally 'materialise'; in any case, he argued, the creation of everything 'in one go' was an even greater leap beyond conventional physics. (Bondi, Gold, and Hoyle came up with their idea, in 1948, after seeing a film called *The Dead of Night*, whose conclusion recapitulated the opening scene.)

The steady-state theory provided a constructive stimulus for more than fifteen years. If a 'steady state' prevailed, then distant regions, even though we see them as they were a long time ago, should, statistically, look *just like* nearby regions – this is a very specific prediction. If remote galaxies look different on average, we can't be living in a steady-state universe.

THE ADVENT OF RADIO ASTRONOMY

Even if our Universe were evolving, changes would be so slow that they would only be manifest over billions of years. To detect an evolutionary trend (or to check whether the universe is really in a steady state) one must probe galaxies so far away that their light set out several billion years ago. Such efforts started as early as the 1950s, using the telescope on Mount Palomar in California (which, with its 200-inch-diameter mirror, was much the largest in the world). The results were inconclusive. Normal galaxies with sufficiently high redshift were not luminous enough to register on photographic plates, even with such a powerful light-gatherer as the 200-inch telescope.

The world's best optical telescopes, in the 1950s, were concentrated in the United States, particularly in California. However, the next observational breakthrough in cosmology came from a quite different technique – radio astronomy. The realisation that some cosmic objects emitted energy in the radio band, not just visible light, opened up a quite new 'window' on the universe. Radio waves from space can pass through clouds, so Europe had no climatic handicap in this new science.

The new subject of radio astronomy was pioneered by physicists who had been involved in radar research during the Second World War. The Cambridge effort was led by Martin Ryle, with encouragement from William Bragg, the Cavendish Professor, and from Jack Radcliffe, an expert in radio-wave propagation and the upper atmosphere.

From the early 1950s to the 1970s, Ryle built a series of increasingly sensitive radio telescopes, based on his own innovative designs. He used these to provide the first maps of the radio sky. He realised, early on, that some of the intense sources of cosmic radio waves were associated with unusual 'active' galaxies (now believed to harbour massive black holes in their centres) which were billions of light years away. Their radio emission could be detected even when they were too far away to be seen with optical telescopes. He was therefore able to probe what the universe was like several billion years ago.

Radio telescopes are amazingly sensitive to very weak signals. Martin Ryle had a nice way of illustrating this. When 'open days' were held at his observatory, each visitor was asked to take a tiny slip of paper from a pile. On it was written: 'in picking this up you have expended more energy than has been received by all the world's radio telescopes since they were built'.

A problem in the early days of radio astronomy was to pin down the exact directions that cosmic 'radio noise' came from. Ryle invented a technique called 'aperture synthesis' which surmounted this problem, enabling him to survey the northern sky and locate several hundred sources. He used his data, very ingeniously, to conclude that our universe was actually evolving, and couldn't be in a steady state.

Ryle didn't know the distances of his radio sources (most had no visible counterpart, so optical astronomers couldn't measure redshifts), but he assumed that the weaker sources were, on average, further away than those giving intenser signals. He counted the numbers with various apparent intensities, and found that there were surprisingly many apparently weak ones – in other words, sources mainly at large distances – compared with the number of stronger and closer ones. It was as though we were in the middle of a huge sphere, several billion light years in radius, and there was a much higher concentration of radio sources near the surface of the sphere than near its centre. This seemed incompatible with a steady-state universe, where the sources must, by hypothesis, belong to similar populations at all times, and therefore at all distances. However, it was quite compatible with an evolving universe. Ryle conjectured that galaxies were more prone to undergo the mysterious outbursts that generated intense radio waves when they were young. If galaxies 'quieten down' as they mature, fewer nearby ones would be detected as radio sources.

Ryle's argument was first put forward in the 1950s, and provoked a noisy (and often ill-tempered) controversy that ran for several years. There was initial scepticism about Ryle's claims because the early radio surveys had been inaccurate – they yielded such a blurred map of the radio sky that two or more separate sources were sometimes counted as one. However, by 1958, when Ryle presented his case for an evolving universe in a major lecture at the Royal Society, the most serious 'bugs' had been dealt with, and his data were reliable; essentially everything he said in that lecture has stood the test of time.

The steady-state theory called some cherished beliefs into question, and offered specific predictions which goaded the observers into attempts to refute it. The theory's originators, an articulate and inventive trio who relished controversy, were effective publicists. Hoyle, in particular, was a brilliant populariser; many younger cosmologists (and I'm among them) owe their original impetus to his books and broadcasts. The confrontation between a steady-state universe and an evolving one accordingly achieved wide public currency. At least this was true in Britain: the voices of Bondi, Gold, and Hoyle failed to carry across the Atlantic, and their theory was never taken very seriously in the United States. But it was the pioneer radio astronomers in Cambridge who were best placed to carry out the crucial radio observations.

Ryle plainly wanted his radio surveys to have a decisive impact on cosmology, which they could only do by refuting the 'steady state'. He had invested years of effort in designing and building new instruments, as well as in the data-gathering itself. No single individual nowadays can master all the necessary techniques; Ryle was an exceptional exemplar of the pioneering radio astronomers who conceived and built novel equipment, and themselves drew fundamental inferences from the data.

Despite Ryle's compelling arguments in 1958 (or so they seem in retrospect), controversy took several more years to die down. To its proponents, the steady-state theory had a deep philosophical appeal – the universe existed, from everlasting to everlasting, in a uniquely self-consistent state. Moreover, if it were correct, every process of importance in the universe would have to be going on somewhere now. On the other hand, in a 'big bang' Universe, some key processes might (or so it then seemed) be for ever inaccessible.

THE RENAISSANCE IN UK OPTICAL ASTRONOMY

A second development – the advent of cheap air travel – has proved equally important in reviving observational astronomy in Cambridge. From the 1960s onward, it became feasible for Cambridge astronomers to use telescopes on good overseas sites: the British climate was no longer an impediment. Optical astronomy is the oldest branch of the subject, but has undergone a renaissance in the last decade. Modern telescopes are enormously more powerful than their predecessors: they can achieve sharper images; new 'solid-state' detectors are fifty times as efficient as photographic plates for detecting faint objects; optical fibre techniques allow spectra to be taken hundreds at a time, rather than just one by one. Cambridge has taken a leading part in these developments.

In 1967, an Institute of Theoretical Astronomy was set up, directed by Fred Hoyle. Its specially-constructed building, funded by the Wolfson Foundation, adjoins the Observatories. (Previously, the theorists had been based in the Faculty of Mathematics, and had been isolated from day-to-day interaction with observers.) In 1972, this theoretical Institute was merged with the Observatories to form the present Institute of Astronomy. The next two decades saw substantial expansion, particularly in observational optical astronomy, and in the development of novel instruments.

Cambridge astronomy received a further boost in 1990, when the Royal Greenwich Observatory (RGO) was relocated to another new building physically linked to our own Observatory. The RGO is not formally part of the University – it is a national institution that provides facilities and expertise for astronomers in all UK universities – but its staff, and the extra resources it brings, are undoubtedly an asset. Moreover, the RGO is better able to fulfil its national role, and foster collaboration with other universities, through benefiting from Cambridge's strong scientific and technical infrastructure.

Research in Cambridge covers almost all branches of astronomy and cosmology – the Sun, planets, stars and galaxies and how they evolve – and the nature of the early Universe. A distinctive feature is its strongly international character, and the close collaboration between theorists and observers whose data is obtained by a variety of techniques – ground-based optical telescopes around the world, infrared and radio telescopes, and spacecraft (including the Hubble Space Telescope, and x-ray observatories).

Large optical telescopes are now revealing, more clearly than radio astronomy ever could, what our Universe was like when it was only about a tenth of its present age. A very long exposure by the Hubble Space Telescope last year revealed our deepest and sharpest picture of the distant universe. An image of just a small patch of sky – a thousandth of the area covered by a full moon – is densely covered with faint smudges of light, each a billion times fainter than any star that can be seen with the unaided eye. But each of these smudges is an entire *galaxy*, tens of thousands of lightyears across, which appears so small and faint because of its huge distance.

What is fascinating about such pictures isn't the record-breaking distance in itself, but the huge span of time that separates us from these remote galaxies. They are being viewed when they have only recently formed. They have not yet settled down into the steadily spinning 'pinwheels', like the beautiful spiral galaxies depicted in most astronomy books. Some consist mainly of glowing diffuse gas that hasn't yet condensed into individual droplets, each destined to become a star. There would not yet have been time for stars to manufacture the chemical elements. These newly formed galaxies would not yet harbour planets, and presumably no life.

MATTERS OF GRAVITY: THE DISCOVERY OF PULSARS

Before anyone recognised the role of supernovae in the 'ecology' of galaxies (described earlier) there had been speculations about how stars explode, and what remnants they may leave behind. The supernova explosion is triggered by the sudden collapse of a heavy star's core when it finally exhausts its nuclear fuel. Most of the material is ejected in the explosion, but a dense 'cinder' remains behind. The central nuclei of atoms are tiny compared to the atoms themselves, whose overall dimensions (and spacing in ordinary solids) are set by the diffuse 'cloud' of surrounding electrons. In a 'neutron star', the nuclei are so close packed that the entire mass of a star could then be squeezed within a radius of 10 kilometres – the volume of a sugarlump would contain a hundred million tonnes of material.

Why neutrons? The nuclei of ordinary atoms are made up of protons and neutrons. For example, helium has a nucleus of two protons (each with a positive charge) plus two neutrons (with no electric charge); iron has twenty-six protons and thirty neutrons. In the laboratory, an isolated neutron decays into a proton and an electron. On the other hand, at high densities the process goes the other way: protons turn into neutrons.

Despite theoretical interest dating back to the 1930s, the idea that supernovae may leave dense remnants remained just a conjecture right until 1968. In that year, an ordinary-looking little star in the middle of the Crab Nebula (the remnant of a supernova that exploded nearly 1,000 years ago) was actually found to be flashing on and off thirty times a second. Nothing but a neutron star could be compact enough to spin so fast without flying apart.

The 'remnant' in the Crab Nebula was not the first neutron star to be discovered. Priority went to Anthony Hewish and Jocelyn Bell, Cambridge radio astronomers, whose discovery of 'pulsars' constituted one of the most remarkable pieces of serendipity in modern science.

Hewish built a special instrument with an important special feature; it was sensitive enough to record *rapid changes* in the intensity of the radiation from distant sources. He was primarily interested in the peculiar distant galaxies that emitted so powerfully in the radio-frequency band. These were the objects whose statistical properties enabled Martin Ryle to strike the first blow against the steady-state theory.

Early radio telescopes, unfortunately, gave a blurred view of the radio sky: they did not reveal, for instance, whether the radio emission came from a small region or from a fuzzier volume. Hewish invented a special technique for diagnosing the size of a radio source. His method exploited the same physical principle which causes stars to twinkle whereas planets do not: starlight is irregularly refracted in the upper atmosphere, but the irregularities are so small that a planet, whose image is an extended disc, covers so many of them that the effect averages out. Hewish discovered that the diffuse gas flowing out from the Sun into interplanetary space, the 'Solar Wind', affects radio waves rather as the upper atmosphere affects visible light.

Hewish found that some of the radio sources did indeed 'scintillate' in the expected fashion. But his research student, Jocelyn Bell (now a professor at the Open University) found variations of a quite distinctive kind – sporadic series of regular pulses each lasting a fraction of a second, which turned out to come from specific points in the sky. A frantic few months of effort ensued. The Cambridge radio-astronomers had to check whether the signals had a terrestrial origin (maybe some secret space project?). Three more of these mysterious sources were soon found, each ticking at a well-defined rate. Could they perhaps be signals from intelligent extraterrestrials? This idea was never taken very seriously, but the sources were jocularly referred to as LGM 1, 2, 3, and 4 (for 'little green men').

When this discovery was announced in the journal *Nature*, even the other astronomers in Cambridge were astonished. Hewish and his colleagues had not shared their excitement with anyone outside a tight-knit group. This concealment annoyed some of us at the time, but in retrospect I think Hewish was no more than prudent. Only a few months elapsed between Jocelyn Bell's first intimations and the actual publication, so nobody's chance of follow-up work was seriously delayed. And for most of those months, Hewish and Bell weren't completely confident that the signals were 'real'. If the sporadic radio pulses had turned out to have a mundane interpretation, or to arise from some fault in their equipment, a premature announcement would not only have been embarrassing, but might have wasted the efforts of many other astronomers who would undoubtedly have followed up any rumour of this kind.

What could these objects be? An ordinary star like the Sun would fly apart if it pulsed or rotated much faster than once per hour. Bodies that turned on and off in a fraction of a second plainly had to be much more compact. Were they white dwarfs, or maybe neutron stars? Were they pulsing or spinning? All these options (and many others) had their advocates. The Cambridge group originally favoured pulsating white dwarfs.

The case for rotating neutron stars was first clearly argued by Thomas Gold (co-inventor of the steady-state cosmology; he had by this time moved to Cornell University in the United States). Neutron stars were expected to form when the cores of heavy stars collapsed, triggering supernova explosions. They would be so small, and have such strong gravity, that they could spin as fast as a thousand 'revs' per second without flying apart. The spin rate would provide a natural stable clock; a 'lighthouse beam', anchored to the star, would send an intense pulse towards us once per revolution.

Only a year later, the debate on the nature of pulsars was settled in Gold's favour. A very fast pulsar was found at the centre of the Crab Nebula, transmitting thirty pulses per second: a white dwarf could neither rotate nor pulsate as fast as that, but such rapid spin was no problem for a neutron star. Moreover, careful timing showed that the pulse rate was gradually slowing down: this was natural if energy stored in the star's spin was being gradually converted into radiation, and into a wind of particles which keep the Crab Nebula shining in blue light.

Pulsars might, had it not been for a 'near miss', have been discovered earlier. In 1964, Hewish and a research student from

Nigeria, Sam Okoye, unknowingly detected the pulsar in the Crab Nebula. They did not actually record pulses, but they discovered that the radio emission from the middle of the Crab Nebula had to come from some concentrated object different from any other source known at that time. By following this up, they might have discovered the pulses. Had history gone that way, the Crab Pulsar would have been the first neutron star to have been discovered. Hewish would still have been co-discoverer of pulsars, but four years earlier, and with Sam Okoye rather than Jocelyn Bell.

Pulsars opened up new prospects in astronomy; the pace of pulsar research has never flagged in the succeeding twenty-five years. Neutron stars are also fascinating to physicists – they exemplify how the cosmos offers a 'cheap' laboratory allowing us to study how matter behaves under extreme conditions that cannot possibly be simulated here on Earth.

BLACK HOLES

Even more extreme are black holes – objects where gravity has overwhelmed all other forces. Reasons for believing in black holes date back nearly seventy years. In 1930, a precocious young Indian, Subrahmanyan Chandrasekhar, enrolled at Trinity College, where he hoped to become one of Eddington's students. During the long voyage to England, he thought about white dwarfs – the dense remnants of stars that can no longer draw on nuclear energy. He reached a startling conclusion. White dwarfs more than 1.4 times as heavy as the Sun couldn't exist: their central pressure could never build up high enough to counteract gravity.

This raised the question as to what happened when heavier stars ran out of fuel. They may, of course, throw off so much material in the course of their evolution that their masses end up below Chandrasekhar's limit for a white dwarf and they fade away gradually. Or the outer layers may be expelled in a supernova explosion, leaving a neutron star, as seems to have happened in the Crab Nebula. But there is also a limit to how heavy a neutron star can be. Stars don't all have the 'prescience' to shed enough gas to bring them safely below this limit, and any stellar remnant more massive than about two or three solar masses would collapse completely when its nuclear energy sources ran out. The supernovae that are triggered by the heaviest stars, those above twenty solar masses, are thought to leave black holes rather than neutron stars.

Black holes are the 'ghosts' of dead massive stars; they have collapsed, cutting themselves off from the rest of our Universe, but leaving a gravitational imprint frozen in the space they have left. Around black holes, space and time behave in highly 'non-intuitive' ways. For intance, time 'stands still' at the surface: an observer hovering there would witness the whole future of the external universe in what, subjectively, seemed quite a short period.

Einstein's equations can only be properly solved for specially simple cases – for instance, collapse of an exactly spherical object, or expansion of a 'model universe' that is completely uniform. Are these reliable guides to what happens in more realistic cases? New mathematical concepts had to be deployed before theorists could analyse collapsing stars or expanding universes that are realistic rather than idealised. Roger Penrose, whose early training had been as a mathematician, was the catalytic figure. In the 1960s he introduced new mathematical techniques which revealed that 'singularities', where the strength of gravity 'goes to infinity', are deeply rooted in the structure of space and time. When the solution to an equation 'blows up' like this – when, as it were, 'smoke pours out of the computer' – it generally means that the theory has broken down, or become in some way inadequate.

In the early days of Einstein's theory, Eddington in Cambridge was its leading champion and expositor. Forty years later, when relativity underwent a renaissance, its most influential Cambridge exponent was Dennis Sciama. It was Sciama who first enthused Penrose, originally a pure mathematician, to work on relativity. Sciama attracted and inspired a steady flow of students, and thereby catalysed many of the key developments in relativity and cosmology. Among these students was Stephen Hawking. Sciama encouraged him to attend Penrose's lectures; these lectures expounded the mathematics that Penrose and Hawking utilised in their joint studies of gravitational collapse. The results of this work were codified in a highly technical book, *The Large-Scale Structure of Spacetime*, which Hawking wrote with George Ellis, another of Sciama's former students.

The renaissance in gravitational research which began in the 1960s was due partly to more powerful mathematical techniques; it was also stimulated by astronomical discoveries. For the first time, astronomers realised that there were places in the Universe – even within our own Galaxy – where relativistic effects could have extraordinary implications.

Most remarkable are the huge black holes which are the 'relic' of the powerful energy sources which Ryle and other pioneering radio astronomers were first to detect. These lurk in the centres of galaxies, where the space of our Universe has been 'punctured' by the accumulation and collapse of large masses to entities described exactly by fairly simple formulae. As Roger Penrose has remarked, 'it is ironic that the astrophysical object which is strangest and least familiar, the black hole, should be the one for which our theoretical picture is most complete'.

THE BIG BANG?

The term 'big bang' was coined by Fred Hoyle in the 1950s, as a derisive description of a theory he never liked. Despite his preference for a 'steady state', Hoyle himself led us towards what has turned out to be one of the strongest pieces of evidence for a big bang.

If the entire universe had once been squeezed hotter than a star, you might wonder whether nuclear reactions could have happened then – indeed, some early proponents of the big-bang theory suspected that the chemical elements were indeed forged in the early universe. However, the expansion turns out to have been too fast to allow carbon, iron, etc. to be built up. But Hoyle and others calculated that there would be enough time for about 25 per cent of the hydrogen to be turned into helium.

The proportion of helium in old stars and nebulae turns out, remarkably, to be just about what is calculated to emerge from the 'big bang'. As a bonus, so are the proportions of lithium and deuterium ('heavy hydrogen') as well. Moreover, these particular elements couldn't be made in stars, even though the stellar nucleogenesis scenario seemed to work so well for carbon, oxygen, etc. Even the oldest objects contain a lot of helium – far more than could have been made in stars; and deuterium is so fragile that it is destroyed rather than created in stars. These considerations therefore vindicate an extrapolation right back to when the Universe was hot enough for nuclear reactions to occur – that's when it was just a *few seconds old*.

BACK TO THE BEGINNING

The grounds for extrapolating back to the stage when the Universe had been expanding for *a few seconds* (when the helium formed) deserve to be taken as seriously as, for instance, ideas about the early history of our Earth, which are based on inferences by geologists and fossil-hunters which are equally indirect (and less quantitative). But can we extrapolate back even further?

To Newton, some features of the Solar System were a mystery. He showed why the planets traced out ellipses. But it was a mystery to him why they were 'set up' with their orbits almost in the same plane, all circling the Sun the same way. In his *Opticks* he writes:

> blind fate could never make all the planets move one and the same way in orbits concentrick . . . Such a wonderful uniformity in the planetary system

must be the effect of providence.

This coplanarity is now understood – it's a natural outcome of the Solar System's origin as a spinning protostellar disc.

Three centuries of progress have given us a vastly broader perspective on the cosmos, and allowed us to extend the causal chain right back from the beginning of the Solar System to the first second of the big bang. But the demarcation between phenomena that are the manifestations or working out of known laws, and those which are mysterious 'initial conditions' still exists, as sharply as it did for Newton.

Not even the boldest theorists can extrapolate back beyond the stage when quantum effects become important for the entire universe. The two great foundations of twentieth-century physics are Einstein's theory of gravity (general relativity) on the one hand, and the quantum uncertainty principle on the other. But there's generally no overlap between these two great concepts. Gravity is so weak that it's negligible on the scale of single molecules, where quantum effects are crucial. Conversely, gravitating systems like planets and stars are so large that quantum effects can be ignored in studying how they move.

But right back at the beginning of the universe, the densities could have been so high that quantum effects could shake the whole universe. This happens at the Planck time 10^{-43} seconds. A theory of quantum gravity is the greatest challenge that confronts fundamental physics today.

EPILOGUE

Astronomy – the exploration of our cosmic environment – is one of the growth points of current science. We are beginning to understand the place of our Earth – indeed our entire Solar System – in a cosmic evolutionary scheme stretching right back to the hot dense fireball in which our Universe began. Black holes, the big bang, and quasars have entered the common vocabulary if not yet the common understanding. New questions about the 'beginning of time', previously entirely speculative, are now within the scope of serious science.

The subject has a widespread intrinsic appeal: in particular, it fascinates young people, and helps to instil an enthusiasm for science. The University's high profile in astronomy helps to boost its international standing; it also enhances Cambridge's appeal to potential students of physical and mathematical science. Not only the United States, but the countries of mainland Europe as well, are now deploying greater resources towards astronomical research. Building on our traditions, and maintaining our standing, is therefore an ever-tougher challenge.

CHAPTER 2

Cambridge's contribution to medical science

MARK W. WEATHERALL

INTRODUCTION

If my task was to outline Cambridge's contribution to medical practice, then it would be most appropriate to concentrate on the University's role in medical education. Medical degrees have been granted at Cambridge for over 700 years, and a steady trickle of students – increasing in the last century to something approaching a flood – has come to Cambridge to be taught the rudiments of medicine, before leaving for London, Edinburgh, or one of the great continental schools where they could see more patients on the wards of their larger hospitals. (In fact at the end of the last century, and again since 1976, Cambridge itself has provided the facilities for students' clinical education.) After several years the students would return and take their *viva voce* examination for the Doctorate of Medicine.

In the past the University's critics held a very poor opinion of the Cambridge-educated physician. Not even the most elevated members of the faculty were immune from criticism: Isaac Pennington, for example, Regius Professor of Physic from 1793 to 1817, began his professorial career as Professor of Chemistry, but quite clearly knew little chemistry and didn't deliver any lectures. Naturally, therefore, he was promoted to the Chair of Medicine, about which he ought to have known a little bit more, being physician to Addenbrooke's Hospital. As he never lectured on medicine either, it is not possible to say whether he knew much about medical science, but one can say that he was a well-respected local figure in the town and University, that he was influential in the Royal College of Physicians in London, and that he was in many ways a typical example of the elite physicians of his day, the last flowering of the era of the gold-headed cane, of physicians who were incredibly rich and extremely well connected.

To write specifically about Cambridge's contribution to medical science is in some ways rather easier – because one can concentrate on specific discoveries and specific people – but it is in some ways harder because before 1870 it is difficult to pinpoint a discovery, an advance, or an invention that was actually made in Cambridge itself. In the last one hundred years there has been an enormous expansion in the number of medical scientists working in Cambridge, and also in the size of Addenbrooke's Hospital, and many important contributions to medical science have been made here. Before 1870, Cambridge's contribution to medical science occurred at one remove, in the education of many doctors and scientists who made pioneering discoveries or advances. The doctors and scientists mentioned below received the formative portion of their medical training in Cambridge: in that sense, Cambridge made them, and they made their contributions to medical science.

CLASSICAL SCHOLARSHIP – JOHN CAIUS

John Caius became one of a select group of men, women, and saints after whom a Cambridge college has been named when he refounded the college that now bears his name (at that time it was known as Gonville Hall). Caius was the leading light of sixteenth-century British medicine: he was President of the Royal College of Physicians of London from 1555 to 1561, again from 1562 to 1564, and once again in 1571, but never held the Regius chair at

Cambridge. At that time, learned medical practice was based on the works of the ancients, particularly Hippocrates, the 'father of medicine', and Galen, the Roman physician whose literary output was only exceeded by his ego. The works of Hippocrates and Galen were felt in many cases to contain the last word on medical subjects, and vast amounts of scholarly time and effort were invested in trying to rediscover, retranslate, and reinterpret these works in as pure a form as possible. Caius was a great Greek scholar, and he was one of the first people in this country to make accurate and insightful translations of some of the great Greek classics in Latin, which was of course the universal learned language of his day.

Besides the fecundity of his classical scholarship, Caius pioneered a new way of disseminating medical knowledge as widely as possible. He also wrote a medical book in English, an undertaking virtually unprecedented for a doctor of his background. Most learned works were written in Latin, but Caius wrote a book in the vernacular for his less-well-educated countrymen who were not fortunate enough to have more than a rudimentary smattering of Latin. He was moved to do this because of the appearance in England of a new disease. One of the major interpretative problems with the works of Hippocrates and Galen was that their authors had lived a long time ago – 1,300 years in the case of Galen and, it was believed, over two millennia for Hippocrates. The physicians of Caius's time were confronted with new diseases that simply could not be found in the compendia of the ancients: one such disease was syphilis, which appeared suddenly around 1500. Caius himself was confronted with a new disease known as 'sweating sickness', which ravaged Britain in an epidemic in the 1550s. Caius's book about this new disease, *A Boke or Counseill against the Disease Commonly Called the Sweate, or Sweatyng Sicknesse*, was the first account of a single disease published in English. To this day medical historians spend a lot of time fruitlessly trying to decide what the 'sweating sickness' actually was from the descriptions that Caius and other doctors wrote.

Caius linked literary ability – the ability to translate and interpret the works of the ancients – with his own observations. He also refounded the college that bears his name, and as it turned out this became rather important for medicine in Cambridge: Caius College became the medical college, at which approximately two-thirds of all those who came to Cambridge to study medicine matriculated. John Caius required that a certain number of the fellows of the college should be medical men (thus ensuring the survival of the medical faculty when other colleges declined to fill their fellowships with any other than mathematicians or divines), and he laid down regulations for the annual performance of dissections, and the delivery of medical lectures. It was to Caius College that William Harvey came in 1597.

ANATOMY – WILLIAM HARVEY AND FRANCIS GLISSON

William Harvey was born in Folkestone, and he came to Caius on a scholarship reserved for people who were born in that town. (At that time many college scholarships and fellowships were reserved for people born in, or connected with, particular parts of the country.) Harvey came to Caius, studied in Cambridge for four to five years, learning natural philosophy and the rudiments of medicine, and then went to study at the leading medical school of the day: Padua. Padua's reputation had been on the rise since the early 1500s – it was at Padua that Vesalius had in 1543 produced the most startling anatomical text of that (or any other) century, *De Fabrica Corporis Humanis* – and by 1600 the fame of the school was at its zenith. It was natural, therefore, that a bright and inquisitive student such as Harvey would gravitate to Padua, and once there, to the anatomists. In Padua Harvey met the anatomist Fabricius d'Aquapendente who, besides lecturing to the students and performing the annual dissections of the bodies of criminals, was investigating the functions of the body's organs. Fabricius's approach, derived from the works of Aristotle, was to study each organ in many different animals and thereby to tease out their essential nature, their purpose, or (in Aristotelian terms) their final cause. According to Fabricius, in order to understand the essence of the lungs, for example, it was necessary to compare their functioning, their size, and their structure in as many different animals as possible.

Harvey applied this approach to the heart. After much painstaking work on animals as diverse as pigs, snakes, and dogs, he found that the major function of the heart was to expel a certain quantity of blood with each beat. To modern ears this sounds obvious, but in the early seventeenth century there was much argument about the relationship of the movements of the heart to those of the blood – whether the movement of blood out of the heart was due to blood being sucked out of the heart, or whether it was due to the heart's force compressing it, or some measure of both. Galen's anatomical works stated that food, once eaten, went to the stomach, whence it was carried to the liver. There blood was created out of the food, and was carried to the heart by the portal vessels. From the heart the blood was distributed throughout the body in every direction, leaving the heart along both the arteries and the veins. Harvey's calculations showed that the heart was pumping out an enormous quantity of blood each day; indeed, that the quantities involved were so great that it was simply not reasonable to believe that it was all consumed at the fingertips, or the feet, or anywhere else in the body. Harvey reasoned that there must be a way for the blood to circulate and return to the heart. Support for his revolutionary idea was supplied by the anatomical researches of the Paduan anatomists, who had discovered valves (which they christened 'little doors') in the veins and the heart. Harvey pointed out that the valves are so designed as to let the blood go out of the heart along the arteries, but not to let the blood go out of the heart along the veins. Thus, he reasoned, the blood must flow back towards the heart along the veins; blood, he believed, was indeed created in the liver, but once it left the heart it circulated until it was consumed in the periphery.

The evidence supplied by the valves was important, because it enabled Harvey to explain an anomaly that no one had ever considered before. If a tourniquet is tied on someone's arm, the veins stand up. The tourniquet doesn't interfere with the blood going out along the arm in the arteries, but it does interfere with venous return along superficial veins, which become congested. People had seen this effect, and had used it for thousands of years to facilitate the letting of blood, a therapeutic procedure designed to remove evil *humors* from the body. What no one had ever noticed was that if you stop blood from travelling along the superficial veins, they should become congested above the point where you tie the tourniquet, if you believe (as everyone did following between Galen and Harvey) that blood travels away from the heart along those veins. Harvey noted that simple observation shows that this was incorrect: the veins became congested below the tourniquet, and hence the blood must be travelling back towards the heart along the veins.

Harvey first expounded his theory of the circulation of the blood in lectures given at the Royal College of Physicians in 1616, and outlined his work at greater length in his book *De Motu Cordis et Sanguinibus in Animalibus*, published in 1628. He claimed, and had demonstrated to the satisfaction of many of his contemporaries, that the blood must circulate. What he couldn't achieve was to show exactly how it did so, because he couldn't see the capillaries that carried blood from the arteries to the veins. This was beyond the capabilities of the magnifying lens of his day, and another century was to pass before the Italian anatomist Malpighi conclusively demonstrated the existence of these capillaries. So although it may seem strange that people continued to believe in Galenic physiology, to many of Harvey's contemporaries the belief that minute capillaries existed to carry blood from the arteries to the veins seemed equally strange. Although the circulation of the blood seems to us now to be an absolutely natural thing – and a major advance in medicine – it took the best part of 200 years to be fully accepted in all parts of western medical practice.

Harvey wasn't the only great Cambridge-educated anatomist of this time. Francis Glisson, Regius Professor of Physic from 1636 until his death, had such broad interests that his twentieth-century successor in the Regius chair, Humphry Rolleston, was moved to dub him 'a philosopher, an anatomist, a physiologist, a good morbid anatomist, an orthopaedic surgeon, and a clinician', the first two of which job descriptions, at least, Glisson would have recognised. Glisson is known to all first-year students of anatomy as a vague historical figure after whom Glisson's capsule, which surrounds the liver, was named. His place in the pantheon of historical medical heroes was really assured, however, by the publication in 1650 of his *Tractatus de Rachitide siue Morbo Puerili*, the first accurate description of rickets. By the middle of the seventeenth century the publication of medical works in the vernacular had become commonplace, and Glisson's monograph was translated into English the following year.

Glisson's tenure of the Regius Professorship of Physic ushered in a golden era in the history of the medical school at Cambridge. Many bright students who normally have gravitated towards theology and thence to the upper echelons of the established church, were dissuaded from doing so by the religious turmoil of the times, culminating in the Glorious Revolution of 1689. Theology's loss was medicine's gain: the number of Cambridge-educated physicians graduating each year reached a peak around the turn of the eighteenth century. This increase was not to last, however. In an Augustan England enjoying unprecedented prosperity, the heat soon ebbed out of religious debates, to be replaced by the new parliamentary party politics. Theology again became an attractive option for the brightest Cambridge students, and mathematics gradually gained the ascendancy as the accepted best method of educating would-be bishops in suitable mental discipline. The status and influence within the University of medicine and its allied subjects waned, and the medical faculty came dangerously close to extinction, a process not helped by the promotion to the Regius chair of a series of uninspiring characters, culminating in the appointment of the much berated Isaac Pennington. Despite this decline in the school's fortunes, Cambridge can still claim to have produced some of the eighteenth century's leading physicians.

THE ENLIGHTENED PHYSICIAN — WILLIAM HEBERDEN

William Heberden the elder was one of the best known of all eighteenth-century physicians. Samuel Johnson called him 'Ultimus Romanorum, the last of the learned physicians', and he was dubbed 'the English Celsus' by the great William Osler. Heberden's only official appointment at Cambridge was as Linacre lecturer from 1734 to 1738, although it is thought that he continued to lecture regularly until he left Cambridge for London in 1748. The Linacre lectureship at St John's College was in fact the oldest medical endowment in the University, founded by the medical humanist Thomas Linacre in 1524. Heberden lectured on *materia medica*, a subject that brought together what today we would call pharmacy, botany and chemistry; his lectures concerned medicines, how to grow or find the plants that they came from, how to dig up or buy the minerals that they were made out of, how to prepare them, and how to use them. Heberden was in many ways a typical Enlightenment physician. Arriving in London in the late 1740s, he undertook an investigation unlike any which had ever been done before: he collected and tabulated the Bills of Mortality to find out exactly how people died in the metropolis. The application of mathematics to the study of large numbers of people was a characteristically Enlightenment project: it became known as statistics. Heberden's study was an early example of what today we might call epidemiology; he counted up all the diseases and casualties recorded over an eighteen-year period, and summarised them in tables. This was the first time that doctors had held anything other than an intuitive idea of what people suffered from, what was common and what was not. The major categories for deaths were 'infants' (the figures were very high, in four figures every year), 'abortion', and 'stillborn': levels of perinatal mortality were clearly very high at this period. Very high numbers were continuously recorded for consumption (galloping tuberculosis) and cough. One also sees epidemics, for example of 'purple [spotted] fever'. Many of Heberden's categories are still seen in mortality tables today, but there are some entries that seem strange to us, such as 'killed by several accidents', or 'suddenly'. These highlight the difficulty of identifying past causes of mortality with the diseases that afflict us today, a problem which besets all attempts to apply epidemiological lessons from the past to plagues and epidemics of the present day. Nevertheless, the actual project of counting the dead and how they died, and of counting ill people and why they are ill, began in the eighteenth century, and Heberden was one of the pioneers.

An interesting side-effect of this project was that it caused many doctors to think about how they actually distinguished the diseases listed in tables such as those produced by Heberden. Many doctors started asking, how do we distinguish what is going on? What is causing these illnesses? Can 'cough', for example, be subdivided into further categories depending on what actually caused it to happen? Another investigative technique pioneered by Enlightenment doctors was the preservation and scrutiny of extended case histories, the stories of patients over decades, the accurate recording of their health and sickness in a chronological account: this was the period during which a recognisably modern 'case history' was created. In the right hands, case histories were very powerful, because in comparing them doctors like Heberden started to notice patterns. These patterns allowed them to distinguish apparently identical morbid conditions which were actually distinct from one another: today Heberden is best remembered in the general history of medicine for his discovery, or description, of angina. He was the first person to distinguish what we would now call ischaemic chest pain as a distinct disease entity. He is also a historical hero to rheumatologists, because he was one of the first doctors to differentiate between different types of joint pain, and to start unpicking the complexities of arthritis.

THE PHYSIOLOGISTS — MICHAEL FOSTER, WALTER GASKELL, AND JOHN LANGLEY

Most of the work for which Heberden is remembered was done in London, rather than in Cambridge. Yet Cambridge's modern reputation as a world-class centre of medical science rests on research done in Cambridge itself. This change began in the early nineteenth century, when the University increasingly came under outside pressure to reform its outmoded teaching practices, its nepotistic and incompetent college and university hierarchies, and its all-too-cosy links with the upper echelons of the professions, Parliament, and the court. Reform was a long and complex process, which resulted in extensive changes in college and university government and the introduction of new subjects (such as natural sciences, moral sciences, and modern languages), whilst retaining many of the old Cambridge traditions. The nineteenth century also saw massive changes in the physical appearance of the University, with much time and money invested in a long series of new museums and laboratories. The centre of this expansion was a site formerly occupied by the Botanic Garden. The corner of this site (bounded by Corn Exchange Street, Pembroke Street, and Free School Lane) was home to a building housing the Professors of Anatomy, Botany, and Chemistry. A new anatomy building was erected in the 1830s; it was called the 'Rotunda' because it was round, to allow the dissection to be made in the middle, and all the students to stand around the outside and watch what was going on. Gradually the building got bigger and bigger. When the Botanic Garden moved between 1848 and 1852 to its new site on Trumpington Road, lecture-rooms and museums replaced the conservatories and flower beds, and the New Museums Site was born. Eventually, at the beginning of this century, the Rotunda was pulled down, a big anatomical museum was put in its place, and all the structures now down the side of Pembroke Street comprised the medical school (now the Department of Zoology). Land on the opposite side of Pembroke Street was purchased by the University from Downing College, resulting in a gradual colonisation of this new Downing Site by the scientific departments over the first three decades of this century.

Nearby, on Trumpington Street, Addenbrooke's Hospital, founded in 1766, was completely rebuilt in the 1860s. John Addenbrooke (1680–1719) was a fellow of St Catharine's College; a rather strange chap, he was interested in black magic and necromancy, and is said to have foretold the hour of his own death. He left £4,500 for the founding of a hospital which, after nearly fifty years of legal wrangling over his will, opened in 1766. It was a small, pleasant building, expanded on several occasions in the early nineteenth century, before being torn down and rebuilt to provide the frontage of today's Judge Institute of Management Studies. As the hospital increased in size and the number of its in-patients grew, there was also a change in the problems that brought those patients in. The arrival of the railway in Cambridge in the 1840s introduced a new series of problems to be dealt with by the hospital's surgeons; before then most of their workload had consisted of accidents to agricultural workers and local labourers, along with minor conditions such as leg ulcers, while the physicians dealt with common medical diseases such as consumption.

The enormous expansion in the number of buildings and students was due to the introduction of new scientific subjects, the most important of which, in terms of Cambridge's contribution to medical science, was physiology. The first Professor of Physiology, appointed to that post in 1883, was Michael Foster. Foster, the son of a Huntingdon surgeon, studied medicine at University College, London. Under the influence of 'Darwin's bulldog', T. H. Huxley, Foster studied experimental physiology, quickly becoming one of the leading British exponents of that new science. When, in the late 1860s, Trinity College was looking for a bright young scientist to enrich its fellowship, Huxley and his

friends (including G. H. Lewes, the long-time companion of the
author George Eliot, who was herself a devotee of the new
sciences of the period) recommended Foster. Foster took up his
post at Trinity in 1870; in 1883 the University recognised his
unique talents as a scientist, teacher, and laboratory administrator
by appointing him to a Chair in Physiology. Foster has always
intrigued historians, for he became the most famous physiologist
of his day without making any of the many discoveries that over
the decades between 1860 and 1914 created a whole new way of
understanding the human body. It is probably true to say that
Foster's lasting contribution to medical science was to create the
world-class physiological laboratory which in the early decades of
this century was regularly featured among the lists of Nobel Prize
winners for physiology and medicine. Between the beginning of
the century and the 1950s, Nobel Prizes were awarded to six
physiologists for work begun, or done entirely in the Cambridge
Physiological Laboratory: A. V. Hill's experiments on the heat
production in muscles, and the role of lactic acid, which gained
him the 1922 Prize, arose out of work done in the first decade of
the century by Walter Fletcher (subsequently to become the first
secretary of the Medical Research Council (MRC)) and Frederick
Gowland Hopkins (of whom more below); two of the laboratory's
alumni, Sir Charles Sherrington and Lord Adrian (graduates of
Caius and Trinity, respectively) were jointly awarded the 1932
Prize for their work on a classic Cambridge problem, the mechanisms of nervous conduction; and in 1964 Andrew Huxley (scion of
the illustrious scientific family) and Alan Hodgkin shared the
Prize for their elucidation of the mechanism of nerve conduction
in the squid giant axon.

Two researchers who might well have won Nobel Prizes or
their equivalent had they been awarded in the nineteenth century
were Foster's 'lieutenants' Walter Gaskell and John Langley.
Gaskell and Langley hold an interesting place in the history of
physiology, for they were the first scientists systematically to
investigate the autonomic nervous system (ANS). We have two

nervous systems working in parallel within our bodies: the somatic nervous system that we use to move, and to control our bodies; and the ANS, that functions most of the time unnoticed, keeping our glands, blood vessels, and other internal organs in order. The ANS comes into its own in dangerous situations, when it is responsible for the 'fight or flight' response of an increased heart rate, sweating, heightened awareness, and so on. Gaskell and Langley were really the first people to work out the anatomy and physiology of the ANS; their broad division of it into three major parts, which each have differing structure and function, has been retained to this day. Indeed Langley's 1921 book *The Autonomic Nervous System* remains the first reference to the subject cited in the most recent (38th) edition of Gray's *Anatomy*, published in 1996.

BIOCHEMISTRY AND NUTRITION — F. G. HOPKINS, MARJORY STEPHENSON, AND ELSIE WIDDOWSON

As we have seen, Foster's Physiological Laboratory was a home of world-class medical research; it also threw off a number of off-shoots. One of the first three lecturers in physiology, appointed in 1883, was Sheridan Lea, whose particular interest was chemical physiology. When Lea had a nervous breakdown in the mid-1890s, his post was taken over by Frederick Gowland Hopkins. Hopkins had been trained in London, at University College and at Guy's Hospital; after several years searching for a suitable replacement for Lea, Foster invited Hopkins to come to Cambridge to teach chemical physiology. Hopkins joined the fellowship at Emmanuel College, but found to his dismay that he was required to teach first-year students basic anatomy, a task which he hated every bit as much as did his pupils. He was promoted to a readership in 1902, and in 1910 Trinity College awarded him a praelectorship, the same position which Foster had held forty years earlier. Hopkins originally pursued his research in the Physiological Laboratory, but in 1914 he was given his own department and his own chair in biochemistry. In 1924 a huge biochemical institute was built for him – the Dunn Institute of Biochemistry, on Tennis Court Road – after his friend and former colleague Walter Fletcher had managed to persuade the trustees of the estate of the philanthropist shipbuilder Sir William Dunn that using their funds to support Hopkins's basic research was the best way to support the advancement of medicine as a whole. Since its opening, the Dunn Institute has ranked among the foremost scientific establishments in the world. A new building for the Department of Biochemistry, on the opposite side of Tennis Court Road (on a site formerly occupied by the rear of Addenbrooke's Hospital), is now nearing completion; it will complement and extend the work that has been done in the Dunn Institute.

Hopkins 'discovered' vitamins. He was doing experiments with rats to determine what dietary components they needed to live and to thrive. He was particularly interested in the amount of protein they needed, and also their requirements for energy; these were the leading priorities of biochemical research throughout Europe and the United States at the turn of the century. He used very carefully purified mixtures of protein, carbohydrate, fat, and minerals, and found that, however much of these dietary components he gave to his rats, they wouldn't grow. He tried the experiment again, this time giving them very tiny quantities of milk in addition to the purified materials, and found that they survived, thrived, and grew. He hypothesised that there must be something in the milk that was necessary for growth in addition to the other food components. Hopkins called them 'accessory food factors', an unwieldy term that was subsequently supplanted by the altogether snappier name 'vitamins'. Hopkins was awarded the 1929 Nobel Prize for his discovery, despite the fact that he wasn't really very interested in vitamins. Vitamin research frustrated Hopkins; he was by training a chemist, and wanted to isolate and purify vitamins chemically. When he failed to achieve this, he got fed up with them and decided to move on to other work. It wasn't until 1932 that a Hungarian scientist, Albert Szent-Györgyi, isolated purified vitamin C, and showed that it was the relatively simple chemical hexuronic (later known as ascorbic) acid. Szent-Györgyi, incidentally, worked in the Dunn Institute for several years prior to his discovery, and always held Hopkins in special esteem for having made it possible for him to continue to pursue his research at a crisis point in his career. Despite this, much of the work done in the Institute wasn't concerned with vitamins at all, but was instead focused on other, more fundamental aspects of biochemistry. Eventually Hopkins's old friend Fletcher and his colleagues at the MRC became so frustrated with the fact that Hopkins wasn't doing any vitamin work that they founded an entirely separate laboratory on the other side of Cambridge for that purpose. Once more Fletcher raided the dwindling coffers of the Dunn trustees, and so the Dunn Nutritional Laboratory was born.

One of the unusual characteristics of Cambridge biochemistry was the fact that the laboratories contained many female scientists, despite the generally rather negative attitude towards women in the University prior to their admission to degrees in 1948. The Dunn Institute of Biochemistry was one of the few places in Cambridge (and one of the few scientific institutes in Britain) where your gender didn't matter as much as the work that you did. Norman Heatley, who worked there in the 1930s, informed me that it was known as 'Hoppy's Dating Agency', because of an outbreak of marriages between biochemists in the 1920s. The freedom and openness of the Institute are reflected in its annual publication *Brighter Biochemistry*, which appeared between 1923 and 1931, for which members of the department penned poems, plays, and other forms of prose. The disappearance of *Brighter Biochemistry* in the 1930s was both cause and effect of a somewhat darker period in the Institute's history, during which it played host to scores of refugee scientists from central Europe, and laboured under the financial stringency to which all enterprises were subject in the depressed years before the Second World War.

Nonetheless first-class science continued to pour out of the Institute. One of its leading stars was Marjory Stephenson. Stephenson, a Fellow of Newnham, was one of the first two women to be elected to the Fellowship of the Royal Society – the highest scientific honour in this country – in 1948, less than twelve months before she died. She was a world-class biochemist, whose life work concerned bacterial metabolism, and ways in which one might interfere with it. Her work formed a solid base on which others built to develop antibacterial drugs; many of the scientists with whom she worked at Cambridge, in particular Ernst Chain and Norman Heatley, were the pioneers of microbiology who in the 1940s rediscovered Alexander Fleming's penicillium mould, and made a 'miracle drug' out of it – penicillin. Today it is difficult to imagine medicine before antibiotics, but as bacterial resistance becomes more common, as reports of intensive-care units closed because of outbreaks of multi-drug resistant bacteria become more regular, we find ourselves increasingly reliant on an understanding of normal bacterial metabolism to stay one step ahead of the germs.

We have already mentioned the circumstances leading to the foundation of the Dunn Nutritional Laboratory. A leading figure in the nutrition research carried out at that laboratory from the 1940s to the 1960s was Elsie Widdowson. In an era when vitamin research was all the rage, Widdowson and her long-time co-worker R. A. McCance were among the few scientists to continue to take seriously the quantitative aspects of diet. They undertook many of their experiments on themselves, their students and their colleagues; their work became particularly useful during and after the Second World War, when rationing was introduced. McCance and Widdowson investigated the minimum requirements for people to remain healthy while rationing was in force. There was a marvellous event at the meeting of the Nutrition Society in 1942 when Widdowson and McCance appeared with their daily diet, which was basically a plate full of potatoes, with a few tiny green leaves around the outside, to provide the vitamins and minerals that they needed.

Self-experimentation was characteristic of medical research at that time. Ethics committees didn't exist in their present form, and often it seemed that the easiest way to do an experiment (particularly in the field of nutrition and other aspects of human physiology) was to try it out on yourself. On one occasion Hopkins's ebullient colleague J. B. S. Haldane, having ingested vast quantities of sodium bicarbonate, disturbed a tranquil punting party on the Cam by informing the professor and his rather surprised guests that he was now excreting the most alkaline urine known to man. Widdowson and McCance nearly had a fatal adventure self-experimenting in 1939, when in the course of one experiment they injected themselves with needles that had been sitting in unsterilised water, and developed a raging septicaemia from which they nearly died.

PATHOLOGY IN ACTION – THOMAS STRANGEWAYS AND PENDRILL VARRIER-JONES

One of the other offshoots of the Physiological Laboratory was the Department of Pathology, which was created in the 1880s for its first professor, Charles Smart Roy, a brilliant scientist whose career was sadly curtailed by morphine addiction. Despite spending much of the last one hundred years in the shadow of their physiological neighbours (metaphorically if not literally), Cambridge pathologists have been responsible for some highly innovative projects to apply medical science to medical practice. Two examples are given here. The first project of interest was created by Thomas Strangeways, who held the post of demonstrator in the Department from 1897 to 1919, and of Huddersfield Lecturer in Special Pathology from 1905 to 1926. Strangeways was interested in a common, debilitating disease about which very little was known, namely, rheumatoid arthritis. He decided that in order to investigate the cause, pathology, and possible treatment of the disease, it was really necessary to establish a research hospital solely devoted to the investigation of patients with rheumatoid arthritis, at which scientific research and bedside medicine could be brought into close proximity. The Professor of Pathology, German Sims Woodhead, approved the plan, Strangeways raised the funds to make it possible, and the Cambridge Research Hospital opened in 1906. It was one of the very first institutions in this country devoted to the type of research that would subsequently become known as clinical science. Despite the strong support and keen interest of senior clinicians such as the cultured Regius Professor of Physic, Clifford Allbutt, the Research Hospital was always in financial difficulties, and although Strangeways struggled on raising funds and doing research (and in the process interesting some of the brightest young pathological and biochemical students in the scientific study of disease), after the First World War it was no longer possible to meet the costs of the nursing and administrative staff required to keep the hospital open. By then it had become apparent that the pathology of rheumatoid arthritis was complex and obscure. Suspecting that a full understanding of rheumatoid arthritis was beyond the research techniques available to him and his colleagues, Strangeways decided to wind down the clinical side of his research, and turned instead to an approach that no one else was using at this time, namely, to look at cell culture – that is, the basic techniques of growing cells in broth, or on petri dishes. Strangeways became convinced that the key to the aetiology of rheumatoid arthritis was at a cellular level, but no one could reliably grow cells *in vitro*. The Cambridge Research Hospital metamorphosed into a laboratory devoted to developing new techniques and applications of cell culture. Sadly Strangeways died in 1927, shortly after this shift was made. The laboratory was renamed the Strangeways Research Laboratory, and for several decades was the leading centre in the world for cell-culture research.

The second innovative project to come out of the Pathological Laboratory during the early decades of this century arose out of a particular research interest of Woodhead and Allbutt: tuberculosis. Tuberculosis was a devastating disease for which there was no cure. So many working days were lost because of tuberculosis that it came to be regarded as an economic challenge to the British government, who set aside £50,000 of public money to set up research establishments to investigate the disease. The body established to administer this money – the Medical Research Committee – was the forerunner of the MRC. It was generally believed that the progress of tuberculosis could be slowed down by open-air treatment in a sanatorium; Thomas Mann's remarkable novel *The Magic Mountain* recounts the fortunes of one young sufferer in such a sanatorium located in Central Europe. The Cambridge pathologists saw the establishment of a healthy, open-air environment as an opportunity both to help sufferers lead useful lives, and to study the natural history of the disease itself. A model village was set up, first of all at Bourn, and then at Papworth, a village north-west of Cambridge, under the inspirational superintendence of Pendrill Varrier-Jones. People with tuberculosis were housed in model cottages; they were encouraged to leave their windows open all the time and always be in the fresh air. Chalets were constructed in which sufferers could do craft work to help the community become as financially self-supporting as possible. In the 1920s a laboratory was built at the village, where they did research into tuberculosis; this was named after Woodhead, who had died in 1921.

The open-air treatment of tuberculosis was made redundant after the Second World War by the discovery of streptomycin, and other anti-tuberculous drugs, and the colony was gradually wound down. But today there remain in Papworth many people who came there before or during the war to recuperate from tuberculosis. Some of the GPs now working in Papworth have patients with unusual medical histories: before the advent of streptomycin it was thought that the best way to prevent the spread of tuberculosis within a sufferer's lung was to starve the bacteria of oxygen. Many patients had the infected lobes of their lungs surgically collapsed, and to keep them collapsed they had plastic balls placed in their chests. Some patients living today still have these balls inside them, and as a party trick can produce them at will.

HI-TECH MEDICINE AND MEDICAL SCIENCE

Today medical science in Cambridge is an enormous enterprise: basic science departments in the centre of town house the descendants of Foster's physiologists and Hopkins's biochemists; at the new Addenbrooke's Hospital, built to the south of Cambridge in the 1970s, the Professor of Surgery Sir Roy Calne has pioneered liver and kidney transplantation techniques (he is also a painter of some renown); Papworth Hospital is a world-class centre of heart and lung transplantation; clinical laboratories at Addenbrooke's are home to scientific research into countless aspects of medical practice, from the prevention of the rejection of transplanted organs, to the genes that predispose to breast cancer, to the epidemiology of common diseases such as stroke, osteoporosis, and ischaemic heart disease. Every year the Annual Report of the University of Cambridge Clinical School is packed with hundreds of brief accounts of medical research of the highest standard; the School regularly achieves the highest approval ratings in the national Research Assessment Exercise, indicating the presence of research of international renown.

New Addenbrooke's site is also home to the Laboratory of Molecular Biology, founded in 1962. Watson and Crick's discovery of the double helix in Cambridge in 1953 (and their celebratory drink at the Eagle that lunchtime), is probably the best-known scientific story of this century. Future historians may well conclude that Watson and Crick's discovery is far and away Cambridge's most important contribution to medical science. Cambridge has been a world centre for molecular biology ever since, and the Laboratory of Molecular Biology has become the third in a series of Cambridge laboratories (in succession to Foster's laboratory and Hopkins's institute) to be home to a succession of Nobel Prize winners, including Max Perutz (1962),

Medical science

Fred Sanger (1958 and 1980, a unique double), Cesar Milstein (1984), and Aaron Klug (1982). With such a concentration of scientific expertise in one place, it is probable that the future contributions that Cambridge will be able to make to medical science will be at least as impressive as those of the past.

FURTHER READING

Brooke, C. N. L. *A History of the University of Cambridge. Volume IV. 1870–1990* (Cambridge: Cambridge University Press, 1993)

Geison, G. L. *Michael Foster and the Cambridge School of Physiology* (Princeton: Princeton University Press, 1978)

Langdon-Brown, W. *Some Chapters in Cambridge Medical History* (Cambridge: Cambridge University Press, 1947)

Needham, J. and E. Baldwin (eds.). *Hopkins and Biochemistry 1861–1947* (Cambridge: W. Heffer & Sons, 1949)

Rolleston, H. D. *The Cambridge Medical School* (Cambridge: W. Heffer & Sons, 1932)

Rook, A. J. (ed.). *Cambridge and its Contribution to Medicine* (London: Wellcome Institute of the History of Medicine, 1971)

Rook, A. J., M. Carleton, and W. G. Cannon. *The History of Addenbrooke's Hospital, Cambridge* (Cambridge: Cambridge University Press, 1991)

Weatherall, M. W. *Gentlemen, Scientists, and Doctors. Medical Education at Cambridge 1800–1940* (Boydell & Brewer, forthcoming)

CHAPTER 3

Cambridge and the study of English

STEFAN COLLINI

I

It may already have struck you that a series of lectures on 'Cambridge Contributions' runs the risk of appearing complacently parochial. Collectively, we may seem to be suggesting that most of the significant developments in the intellectual history of the world originated within this University; the result may all too easily end up sounding like a whole fanfare of blowing our own trumpets. I have no wish to contribute to this effect, and it may help if I say from the outset that the reasons why a particular place gets a reputation for having made such an important intellectual contribution to a given discipline often have a great deal to do with luck and timing, with the state of the discipline at that particular moment, and even, if we are realistic about it, with the political and economic power of the country in question. If a lecture-series such as this were to be given in one hundred years' time from now, I suspect that the person lecturing on the study of English would probably have to say that, in the English-speaking world at the beginning of the twenty-*first* century, the really significant and influential intellectual developments were much more likely to be found in the United States than in Britain. In terms of the work associated with individual universities, the imaginary future lecturer would be more likely to concentrate on the influence *on* the people teaching English at Cambridge *of* figures from Yale or Johns Hopkins or Berkeley rather than assuming that the balance of intellectual trade was the other way round. So, these things change with historical circumstances, and there is no divine spirit of intellectual innovation which has made its permanent home within a quarter-mile radius of the Lady Mitchell Hall.

Nonetheless, it is the case that certain developments associated with the study of English in Cambridge, beginning in the 1920s and 1930s, did for various reasons make a quite decisive contribution to the way people all over the English-speaking world thought about the activity of literary criticism and undertook the study and the teaching of English. My aim is to try to give you some sense of what that contribution was and why it made a difference. But of course, if you are to have a sense of the difference it made you really need some understanding of the way things were before it had this impact. I shall, therefore, begin by taking a somewhat longer historical view, going back to the middle of the nineteenth century, to give you a sense of how the whole activity of studying English developed to the point where the intellectual changes associated with so-called 'Cambridge English' came about.

II

At first sight, we are inclined to think that the activity of literary criticism has always existed. After all, we all engage in it, in a primitive form, whenever we are asked 'what did you think of such and such a book?', 'why did you like or dislike it?', and so on. And equally, we are prone to think that the academic 'disciplines' corresponding to this and similar activities must always have had their place among the range of university studies. This is part of the more general tendency to take our present state of affairs for granted, and to assume that current arrangements are 'natural'. So, to help you begin to escape from these assumptions, I want to start by getting you to imagine how things might have developed differently. It was not inevitable that university departments and courses focusing on the literary-critical study of works of English literature should have developed in the way they have, and so I want you in your imaginations to take yourselves back to the point where these did not exist and to think how things might have developed differently.

Well, the first very large difference might have been that people did not think that literature was something particularly worth studying or paying systematic attention to at all. Literature might have continued to be seen, as it certainly has been in many times and places, as essentially an adornment, a source of pleasure or distraction – something, like a well-stocked drinks' cupboard or a conversation with friends, that helps us get through the day. On this view, literature is not something that repays serious intellectual investigation, and, indeed, it may even be something that is rather bad for people. (Plato is probably the most famous and earliest of those who argued that literature should be kept out of the hands of the young, precisely because of its power to distract and excite them.) So it is not to be taken for granted that, even once the behaviour of the natural world or of parts of the human world had come to be thought proper objects of disciplined enquiry, people would necessarily think that literature was an equally serious and valid subject to study systematically. That, in fact, was a conception that only really gained general support in the course of the nineteenth century.

My second hypothetical possibility is to imagine that, even if people had come to think literature worthy of such systematic attention, it was not inevitable that this activity would take place in universities. We tend to take for granted that much of the serious intellectual life of our societies is going to be carried on in universities, but, once again, that is a very recent development. In Britain, for example, until after the middle of the nineteenth century, there were really only two universities and what they

mostly did was either to provide a little finishing-school experience for the sons (but not the daughters) of the rich, or else to serve
as seminaries for Anglican clergymen. They were not serious
centres of the nation's intellectual life in the way that we like to
think they are now. And something similar was true in most of the
other countries of the western world. The growth of universities –
the increased number of them, the rise in the size of the student
population, the expansion in the number of subjects studied there,
and the accompanying belief that if there is a subject that is in some
way worth systematic attention, then universities are where it
should be pursued – all this is something that has really only
grown up with the development of the research ideal of scholarship from the late nineteenth century onwards.

My third way of getting you to imagine how things might have
been different is to suggest that, even if you thought literature was
important and even if you thought the university was the place to
study it, there might still be no reason to confine it to *English*
literature. After all, if you think about it, it would seem a bit funny
to have a department which was called something like 'English
economics' or 'English philosophy', so why do we have one called
'English literature'? Why did it not happen that 'literature' was the
subject of the discipline rather than 'English'? There are, in fact,
rather complicated reasons why this has happened in the way it
has: obviously the whole question of linguistic competence is
relevant, though I think the power of sheer cultural nationalism
is not to be under-estimated either. But in any event, the fact is,
once more, that this is quite a recent development. For much the
greater part of European history, the Classics – that's to say,
Greek and Roman literature – were thought to be the only serious
form of literature, and as recently as 150 years ago it would have
been thought a pretty shocking idea to suggest that the Classics
might be displaced by the study of literature which is in the
language you happen to speak and hence is pretty easy to read. It
was not at all inevitable that the enthusiasm for the study of
literature should issue in the subject of *English* literature.

My fourth hypothetical possibility is really the reverse of the previous one, and that is to suggest that even if you felt that it was your own culture and society you wanted to study, there would be no reason to confine it to English *literature*. We have become pretty familiar with courses and departments of 'area studies', as in 'Latin-American studies' or 'South Asian studies', and one might similarly assume that 'English studies' would have grown up, involving the study of the sociology, the politics, the history, as well as the literature of this society. Once again, the actual historical explanation of why the subject did not take this form is quite complex, but clearly a major role was played by the conviction that there was something quite special about that body of writing we have come to refer to as 'literature', and that its study required a technique and a kind of approach that was different from the sorts of skills that might be involved in dealing with other material. And although in recent decades there's been something of a move away from this conviction, as a result of which there is now an increasing tendency to treat literary texts as like the other cultural products of a particular society, still by far the most common form taken by courses and departments around the world is that of 'English literature'.

My fifth and final hypothetical possibility is of a slightly different kind. As I said earlier, we have come to take for granted that at the heart of the study of English is the activity of 'literary criticism'. Close attention to the verbal detail of a poem or a play, the analysis of how those details work by examining the images and metaphors or identifying the tone and register and so on, and then the attempt to harness these details to an overall 'interpretation' and assessment of the 'value' of the literary work – this we have come to think of as the essential activity involved in literary study. But again, it might not have developed in this way, and, as I'll explain in a moment, it was not like that as recently as the beginning of the twentieth century. There are, after all, other ways in which you could engage with that stack of old books we have come to refer to as 'English literature'. You might, for example, treat them historically, you might go to them for the kind of information they provide about past societies, you might treat them purely as sources of data about the development of the language, and so on. The idea that 'criticism' is a genuinely respectable intellectual activity which can be studied and taught, and that it is an activity of considerable social and even moral value, rests upon several distinctive intellectual assumptions and cultural circumstances, and the idea that it should be at the centre of a university course in English would, at the beginning of this century, have seemed really quite a revolutionary pedagogical idea.

So, in each of those ways, things might have developed differently, and the study of English might not have evolved in the way it has. I would like you to hold those possibilities in your minds as I now give you a very brief indication of how it actually *did* develop before going on to concentrate on the difference that work associated with Cambridge made to it. The place to start is with the huge expansion of universities that took place across Western Europe and the United States beginning in the last few decades of the nineteenth century. In Britain new universities were founded in the great cities that had developed as a consequence of the Industrial Revolution, in places such as Leeds, Manchester, Birmingham, and Liverpool, and a central element in this expansion was the introduction of a range of new subjects alongside the venerable disciplines of mathematics and Classics – subjects such as history, natural science, modern languages, and so on.

As part of this general expansion, there was considerable debate about whether English literature, too, should be established as one of the new academic subjects. It was recognised that with the beginnings of a more democratic system of education at school and university level – universal elementary education had been introduced for the first time by the Act of 1870 – it would be unrealistic to confine literary education to the study of Greek and Latin. But if the study of English were in any sense to take the place of the Classics, then it had to have one characteristic which the study of the Classics had long had in practice: that is, it had to provide the kind of training of the mind that the mental gymnastics of the minute study of the ancient languages was thought to involve, or, in other words, that it had to be *hard*. It was widely felt that there was no point in just letting people go and read books they'd read anyway (and, worse still, might even enjoy); what was particularly needed, as in all educational systems, was something which could be properly *examined*, and for this purpose it was no good encouraging a lot of subjective response about 'the beauties of literature' or, as one professor put it in opposing the setting-up of an English course at Oxford in the 1880s, not just 'a lot of smart chatter about Shelley'. Classics had suited this purpose wonderfully well, and there is a lot of testimony from those who studied Classics in the nineteenth century that they had little sense of the value or, still less, the pleasure to be derived from the works they were studying, and instead saw them as daunting assemblages of gerunds, ablative absolutes, and irregular verbs. With this model in mind, another commentator on the discussions about the status of English in the late nineteenth century observed that it would obviously help the case for the new subject 'if English could be made to appear a dead language'.

One of the earliest forms of the study of English in some ways did just this. As part of the great contribution to the intellectual life of the western world in the late-nineteenth century made by the model of German historical scholarship, the most prestigious form of the study of language was comparative philology, that is, the study of how modern forms of vocabulary and syntax and so on evolved out of earlier and related groups of languages. This was the dominant model when the first university courses in English were being established in the 1880s and 1890s, and for this reason those courses contained large components of the study of old languages from which modern English was descended, such as Gaelic, Old Norse, and above all, of course, Anglo-Saxon. And even where the subject matter was ostensibly post-medieval literature in English, the attention at this time, under the influence of the Germanic model of historical scholarship, was very much concentrated on knowledge *about* literature, including biographical knowledge about the author and the period, and a lot of information about different editions and changes to the texts (again attempting to emulate the prestige of the Classics where the highest esteem was reserved for the scholarly work of emending and editing texts). Thus, in the earliest courses of English we find that concentration was largely on the historical development of the language and on the historical study of information about the books and their authors.

To get some sense of what was involved, you can conduct another little thought experiment, and imagine yourselves as students not in 1996 but in 1896. If you had come to university to do a course in English literature then, what would you have been doing? How would it have been different from what you might do now? Well, of course, one very big difference in many of the major universities of 1896 is that half of you would not be here, because although the number of institutions that were fully open to women was expanding during that period, it was still very restricted, and Cambridge, as you probably know, was not one of them, even though two women's colleges had been founded by this date. (It is important to remember that the growth of English as an academic subject in the twentieth century was closely bound up with the increased educational opportunities for women, since a majority of the students of the subject, and a higher proportion of its teachers than in most other disciplines, have been women.) But if you managed to gain entrance to a university offering a course in English, what would you have been studying? One part of the answer that I have already mentioned is that you would almost certainly have been learning a good deal of Anglo-Saxon. But, more strikingly still, you probably would not have been reading any recent modern literature. I don't just mean that you wouldn't have been reading last week's literature, the books that aren't even yet in paperback. I mean you wouldn't have been reading literature from the last hundred years. When you look at the syllabuses for the study of English at the end of the nineteenth century, what you find is that, even leaving aside the heavy linguistic and philological component, the study of literature usually stops somewhere round about the eighteenth century, and that it has a strong emphasis on medieval and renaissance literature. And one other major difference I think you would find is that whereas now your professors all try and bully you into reading the texts for yourselves (and you, on the whole, try to pretend that you have), in 1896 that was not the central expectation. It's very striking how little attention was paid in the teaching of English to reading the text in that close, detailed, analytical way that we now take for granted. Instead, you would much more often have been expected to write your essays and term papers on the life of the author or on certain facts about the production and editing of the books, and so on.

Although this, roughly, is what being a student of English would have involved in English universities a century ago, it is important to remember that Cambridge was not one of those universities because Cambridge had up to that point resisted the idea of establishing a separate degree in English literature. And this is where I think my earlier point about luck and timing comes in, because the fact that Cambridge did not set up a degree in English when the model was Germanic, historical, and philological meant that the study of English at Cambridge escaped having those emphases built into it from the start. The English Tripos at Cambridge was not established until 1917, a date that was in the middle of an episode of rather larger historical significance than the founding of the Cambridge English course – namely, the First World War. And the First World War made two relevant differences here. The first was that, as always happens during a major war, a lot of people were away at the front, very few people were left minding the shop, and that meant there were fewer obstacles to change. I suspect that most people who have tried to reform a syllabus in their own university must, in their callous moments, fantasise about how the outbreak of a war would help get all the obstructive people out of the way and thus make it easier to get the change accepted, and that's essentially what happened in Cambridge. The other difference that the war made was that it aroused a great deal of anti-German feeling. As you probably know, there was a good deal of smashing of shop fronts which had German names and things like that during the First World War, and the academic equivalent of this behaviour involved a strong reaction against the model of German scholarship, especially against that kind of philological study which emphasised the common Teutonic roots of English and German culture. A course of study set up under these circumstances was more likely, there- fore, to have a larger concentration on the native literature and to pay less attention to the historical development of the language.

So, the new course in English was taught for the first time in 1919, and after a general university reorganisation in 1926 the two-part Tripos was established in something resembling its modern form. But so far I have largely emphasised the negative side of the story, what was *not* in the new course of study or what it escaped by being set up in 1917 rather than in, say, 1887. But what of the distinctive ideas about the study of English which actually did inform this new course of study?

III

I should say here that anyone talking about the contribution of 'Cambridge English' has to perform a slight conjuring trick at this point, because in some ways the intellectually most important figure in the change I'm about to describe actually wasn't at Cambridge at all, and that is T. S. Eliot. Although Eliot later became an Honorary Fellow of Magdalene College, he was an American who came to England as a young man and who wrote his poetry and criticism while working in London, not attached to any academic institution at all. But his critical ideas, developed in the late 1910s and early 1920s, were a crucial part of the intellectual ammunition for the 'revolution' in English studies (as it's often described) that was brought about above all in Cambridge in the 1920s and 1930s. It would take another chapter to describe those ideas, but as a piece of shorthand one could say that both his poetry and his criticism involved a repudiation of the aesthetics of Romanticism that had continued to dominate the discussion, especially the journalistic discussion, of literature in England right up to 1914. In particular, Eliot insisted that criticism was to be seen as an activity of the intelligence and not as a form of dilettantish or gushing 'appreciation'. At the same time, he insisted that 'criticism' involved a close engagement with the verbal organisation of the text itself, and not simply a familiarity with a lot of biographical and historical information about the author and the period. To the young critics and students of literature in the 1920s Eliot represented an attractive kind of 'rigour' and 'discipline', and several of them were eager to introduce these Eliotic qualities into the new English Tripos at Cambridge.

The two most important of these figures, and the two whom I shall concentrate on here, were I. A. Richards (1893–1979) and F. R. Leavis (1895–1978). But I should make clear that there have been a lot of other influential critics and scholars associated with 'Cambridge English' since that period, beginning with the figure whom some would regard as the most gifted *critic*, in the narrowest sense of the term, of the twentieth century, William Empson (who was Richards's pupil). Thereafter, the list of famous names could include L. C. Knights, Muriel Bradbrook, Basil Willey, Raymond Williams, Frank Kermode, and many others. But since I am here dealing only with the change that 'Cambridge English' was supposed to have brought about in the situation I outlined a moment ago, I shall confine myself to the work of the two earliest, and in some ways most influential, of these figures, that is to say, Richards and Leavis.

When considering Richards's contribution, it is important to recall that as a student at Cambridge in the years 1911 to 1915 he studied what were then called the 'Moral Sciences', that is, essentially philosophy but also a substantial element of what would now be studied separately as psychology. Richards went on to do some preliminary medical study with the idea of becoming a psychoanalyst, then a very novel occupation in Britain, but, largely through a series of accidents, he was recruited to teach for the new English Tripos when it began in 1919. But from his intellectual background Richards brought two distinctive concerns to the teaching of English. First of all, he had been strongly attracted to the analytical method of philosophy practised by one of his teachers, G. E. Moore. Moore characteristically asked of all general statements: 'what do we mean, what *can* we mean, by saying, for example, that this is good or this is beautiful?' In other words, Moore was not content to accept the general terms used in such statements at face value, but probed remorselessly for what they 'really' meant, regardless of what we thought we were saying in saying them. And Moore's tendency when dealing with moral or aesthetic terms was always to look to something in the consciousness of the speaker rather than to some alleged property of the outside world when analysing such general statements. And this reinforced in Richards the second unusual concern he brought from his earlier studies, namely a concern with the functioning of human psychology, and especially with its basis in our physiology or nervous system. In this way, Richards had become particularly interested in what you might call the psychology of literary response. What happens in a reader's mind, ultimately in a reader's nervous system, when reading one kind of book rather than another? And he came to believe that one of the things that distinguished really good literature – for example, a really brilliant, tightly structured lyric poem – from inferior literature was something to do with the degree of nervous excitement, the kind of integrated nervous response that it produced in a reader by comparison to the rather low-level or slack or diffuse response which lower-grade or less-concentrated writing was capable of producing.

When Richards started to lecture for the new English Tripos, he tried to work out these ideas, and this led to the writing of the first of the two books from this period for which he is best remembered, *Principles of Literary Criticism*, published in 1924. Richards was a hugely popular lecturer, with something of a cult following, and his other famous book, simply called *Practical Criticism*, published in 1929, grew out of what was essentially an experiment in applied psychology that Richards conducted with his lecture-audiences in the 1920s. What Richards did was to give out to his audience a short poem from which all the identifying marks had been removed – no author's name, no date, no surrounding textual material; just this short text of a poem. And he asked his audience to record before the next week their analysis of, and their response to, this poem. He then collected these responses, and devoted the following lecture to an analysis of them. In the book, he prints each of these anonymous poems, and then follows them with his account of these responses.

One of the things Richards was doing in this little experiment was testing how carefully people could actually *read*. If the poems had been identified, he believed, students would immediately have had all kinds of preconceptions and associations about them which might well block or distract them from paying attention to (in what became a celebrated phrase about the practice of 'practical criticism') 'the words on the page'. Richards wanted his students to have to record their responses without knowing whether the poem was by one of the great names of English literature or by one of one's contemporaries in the same poetry-writing evening class. And, of course, what Richards demonstrated was that all kinds of irrelevant associations and what he called 'stock responses' prevented readers from actually reading what was before them: certain preconceptions about what a poem on a given subject *ought* to be about, certain autobiographical associations set off by a particular phrase, and many other kinds of more straightforward misreading. Richards's conclusion was that people need to be trained to read, to really read, and that the artificial conditions of the practical criticism exercise were an excellent way of encouraging this close, attentive kind of reading.

But Richards didn't think this was only, or even primarily, important for the study of English, and here, I think, we touch on one of the reasons why he was such a charismatic and influential figure at the time. One of the things that was so compelling about Richards was that he believed that everybody, and not just students, needed a kind of inoculation against the dangers of mass communication. In a modern society, in Richards's view, we are increasingly bombarded with messages from the social environment around us – by advertising, by political slogans, by broadcasting, and so on. Our critical response is correspondingly dulled, and we buy things we don't want, vote for things we don't understand, and so on. Richards believed that we therefore needed a training in critical discrimination, needed it as citizens and not just as students. And for this to be effective, we needed to know what it was like to have the more complex and more integrated response provoked by the experience of reading, really reading, a fine piece of literature. These classes in practical criticism were, in his view, how the study of English actually made a contribution that was ultimately of some social and political consequence. It was literary *criticism*, not mere acquaintance with the kind of information provided by literary *history*, that could discipline readers to be capable of resisting the numbing power of modern 'mass' communications (both Richards and Leavis tended to have a remorselessly pessimistic view of the new developments in newspapers and broadcasting, hardly registering their positive potentialities).

His model of using unidentified extracts for these exercises was later also applied to prose, although of course there is usually a little more difficulty in extracting from, say, a novel whereas a short poem is much easier to deal with. In any event, the exercise of 'practical criticism' became, from the mid-1920s, an absolutely staple part of the course in English at Cambridge, and it still is. Students who study English here do an exercise in practical criticism every week of their undergraduate years, and it is an obligatory paper in their final-year exams. And although most other universities around the English-speaking world did not make it quite so central to their syllabus, a critical training of a broadly similar kind did become characteristic of the university study of English almost everywhere.

Of course, as a pedagogical device 'practical criticism' had certain obvious practical advantages. For one thing, it didn't require a lot of books and scholarly resources; for another, it didn't expect the students already to have extensive historical or cultural knowledge. In both these ways, it was well suited to the teaching situation in an age of mass higher education. At the same time, it encouraged students to refine and articulate their own critical response to a piece of literature, and it suggested to them that the adequacy of their response was something of an indicator of the adequacy of their capacity for experience. And yet in some ways Richards's immense influence on literary studies was curious because it would only be a slight exaggeration of the truth to say that he wasn't primarily interested in works of English literature: he was *primarily* interested in ways of training and measuring critical response in the reader, and the great works of English literature just happened to be the most useful pieces of equipment for conducting experiments on this topic and for providing this training. Needless to say, a fuller account of Richards's career would have to do justice to the many other sides of his work, including his interest in developing 'Basic English' as a tool of international communication, and his role in shaping the 'general studies' curriculum at Harvard after 1945. But through his work in the 1920s, as well as through the work of his pupil William Empson, whom I mentioned earlier, who displayed an unparalleled virtuosity in close critical analysis of this kind, Richards's name will forever be associated with 'practical criticism'.

No one would ever dream of saying about the second figure I shall discuss here, F. R. Leavis, that he wasn't primarily interested in the works of English literature themselves, for one of the things that characterised Leavis from his earliest work in the late-1920s right through to his final works in the mid-1970s was his intense, passionate commitment to the value of great literature, particularly English literature. Leavis did not share the kind of scientific curiosity about the psychology of human response that partly animated Richards's work; instead, his was essentially a moral preoccupation, a concern with what the serious business of living a human life should involve. Despite these and other differences, however, and despite the fact that they later fell out with each other (as Leavis, a notoriously difficult and cantankerous figure, fell out with almost everyone with whom he was ever associated), in the 1920s and early 1930s Richards and Leavis shared a concern with the ways in which the really responsive reading of literature could help to counter the deadening and corrupting effect of living in a modern commercialised society.

Leavis was a Cambridge figure in a quite exceptional sense, in that he was born in Cambridge, went to school in Cambridge, and spent his entire adult life in Cambridge (apart from a period of service as a hospital orderly in the First World War). But he had a somewhat chequered academic career in Cambridge (one of the sources of his bitterness and difficulty as a colleague), and for some years made a living as a freelance teacher in the colleges before, relatively late in life, securing a full academic appointment in the University. Nevertheless, through his teaching, through his numerous books, and through the journal *Scrutiny* that he effectively edited throughout its existence (1932–53), Leavis came to be identified as the central representative of 'Cambridge English', even though the running of English at Cambridge was largely in the hands of colleagues who only partly shared his convictions and his critical approach or were in some cases downright hostile to them.

Leavis was clearly an inspiring, if also in some ways intimidating, teacher, and he imbued several generations of students with the sense that the study and teaching of English was simply the most valuable activity one could pursue in life. This meant that he and his pupils could sometimes seem rather patronising or dismissive towards lesser activities, such as studying natural science or running a university, since those activities so clearly did not require that each individual confront, in as intense and focused a way as possible, the meaning and value of a human life. And this same fiercely judgemental quality could be deployed, with even more venom, on those who, though they were institutionally charged with the task of teaching English, revealed by their comments and their tastes that they lacked the fundamental intellectual and emotional resources required to engage with great literature at a serious level. In this respect, the 'Cambridge contribution' has sometimes seemed to be the widespread conviction among colleagues in other disciplines that the kinds of people who teach English are impossibly posturing, fractious, and incapable of sensible cooperative action.

In terms of the actual study of English, there were perhaps three main things that became associated with Leavis's name. First of all, he developed in a more explicit and systematic way the implications of T. S. Eliot's critical practice. Following Eliot, Leavis insisted that sensibility and intelligence were not separate things in one's experience of literature; they were inseparable aspects of a single response. What is usually seen as a simple failure of sensibility, such as not picking up the emotional force of a piece of writing or not properly identifying its tone, always involved a failure of intelligence. To become aware of and accurately to identify an experience is partly a matter of reflection and analysis, not simply a spontaneous overflow of emotion, and Leavis insisted this was particularly true of that intensification of experience that is involved in reading a powerful work of literature. This was one of the ways Leavis and his followers attempted to rebut the charge that English wasn't a 'real' subject because it only involved mere subjective response to the 'beauties' of literature: on the contrary, they insisted, an education in literary criticism is a real discipline, one which uniquely educates the intelligence and the sensibility together. And as part of this training, Leavis tried to prevent his students from hiding behind the opinions of other critics or behind historical information about an author: he always put them on the spot by forcing them to articulate and justify their own response and evaluation (this was one of the things that made him intimidating).

The second characteristic of Leavis's contribution is closely related to the first, and that is his insistence that only English, among the modern academic disciplines, involves a training in human judgement and evaluation. He complained that in the modern professionalised world of scholarship and research, all the other disciplines, even philosophy, had become merely technical and had abandoned the ancient enquiry into what makes a human life worth living. But in responding to a work of literature – in judging the nature of a poet's experience or the portrayal of a character in a novel – such fundamental judgements about life are inescapable. Leavis did not believe that we wheel each work of literature up to some given, pre-existing scale of values, some variant of the Ten Commandments, and then ask of the work how many of these values it upholds or denies. Rather, he argued, what happens when reading literature is that the more intensely you realise what a passage or a novel or a poem is about and what it is doing, the more you will find you have *already* become involved in judging it, judging how persuasive or admirable or whatever it is. To characterise fully *is* to evaluate. And one of the things an education in English can do, he claimed, was to train us to articulate and justify the judgements that are always implicit in such characterisations. And he particularly emphasised that the judgement *is* always implicit, and that those scholars who claim otherwise, and who claim to give an entirely neutral account of how a poem or novel 'works', are simply trying to hide from the responsibility to say why that piece of literature matters, what picture of life it represents, and how defensible or otherwise that picture is. Mere scholarly knowledge *about* literature was, for Leavis, a sham and an evasion.

And the third thing that Leavis stood for was a very strong emphasis on literature as the source of continuity with an earlier stage of English life and history. That is to say, Leavis thought that industrialisation and the social developments associated with it had made a drastic change in the nature of life as most people experienced it, a change largely for the worse in his view. And so thorough had this change been that the only way now of reconnecting with what the real lived experience of life was like in the pre-industrial periods of English history was through the great masterpieces of English literature. Leavis tended to see these works as distillations, distillations of the most intensely experienced life available at the time they were written, and the only way for us now to make any kind of effective, animating contact with the fundamentally healthier, more fully human, life he believed to have been lived in those periods was through the fully responsive reading of these works. One of the consequences of Leavis's belief in this continuity was that he thought it was most important to study the great masterpieces of *English* literature, precisely because there was an actual historical continuity with this past society. By contrast, he was hostile to the 'great books' courses which at the time were particularly common in the United States; he saw no value in a supermarket of literary works, a selection promiscuously gathered from different periods and cultures. The great task that needed doing was, as it were, to re-educate our minds and sensibilities by reconnecting us to this more valuable, this more genuinely human, form of the life of our own community, which was in danger of being entirely lost in a modern commercial society.

Now, if you put together the things that I've been saying about Leavis, and especially if you combine them with his passionate and intransigent temperament, I think you can see why he communicated a terrifically strong sense of mission to so many of those who came into direct contact with him. One of the most characteristic forms taken by this missionary zeal was the choice of school-teaching as a career, and Leavis was particularly influential on the teaching of English in those sixth forms that flourished in grammar schools in Britain between the 1944 Education Act and the introduction of compulsory 'comprehensive' education in the 1970s. And from the 1950s onwards, Leavis became a prominent and controversial figure in British culture more generally – to many people he is probably best known for his savage attack in 1962 on C. P. Snow's *Two Cultures* thesis. But within English studies some of his most striking impact was to be felt in other countries, with English departments from Melbourne to Manitoba being divided by fierce pro- and anti-Leavis controversies. Indeed, in some ways, his legacy still seems more vital, less a museum exhibit, in certain parts of the former British empire than in Britain itself.

In Britain, here largely responding to developments in the United States although importing some ideas directly from continental Europe, the past three decades have seen a strong reaction against the approach to the study of literature discussed in this lecture, the new styles of work being grouped together, unhelpfully, under the label 'theory'. The belief shared by Richards and Leavis and crucial to their claims for the role of English, the belief that the kind of writing we call 'literature' is distinctive and is distinctively powerful and valuable, has been very strongly attacked. At the same time, their notion of the reader's direct engagement with the text, with 'the words on the page', has also been disputed; in its place there has been more emphasis upon the historical forces which shape the text and upon the sociological and ideological determinants of the act of reading. The objections made to the work of Richards and Leavis and their followers in the name of these new theoretical approaches may not always be well grounded, and may not always be entirely accurate or fair in their account of their targets, but those are large (and delicate) issues that would need to be addressed at another time.

IV

'Cambridge English' came to signify more than simply the fact of certain work being done at Cambridge, just as, at other times in other disciplines, 'Oxford philosophy' or 'Chicago economics' signalled something more — a set of values or assumptions associated with a particular style of work. In this sense, the heyday of 'Cambridge English' lasted from the 1920s until the 1960s. It was never as influential in the United States as in Britain, although in the period after 1945 the so-called New Criticism, a methodological cousin that also owed a good deal to Eliot and something to Richards, came to represent the most widespread form of 'close reading' in the study and, especially, the teaching of English in American universities. From the 1960s onwards, the example of the Cambridge critics was increasingly challenged, repudiated, and, ultimately, ignored (always the main form of inter-generational intellectual change). And in the course of time, the kind of work done within the English Faculty itself at Cambridge — which had anyway always been more diverse than the labels and slogans allowed for — ceased to represent a distinctive approach. English at Cambridge in the 1990s has become a microcosm of the world of literary studies in Britain as a whole in a way that had not really been true in, say, the 1930s.

The danger involved in trying to isolate the 'contribution' of any one institution or group to the development of an intellectual field is, of course, exaggeration. Invariably, the historical story is a complex one, involving many social, economic, and other extra-intellectual factors. Richards, Leavis, their associates, and their pupils, were indeed influential, but this was the result of a coincidence of circumstances for which they were largely not responsible — the widespread perception of the threat to 'culture' from 'mass society', the expansion of universities, the increased professionalisation and self-conscious 'value-neutrality' of other disciplines, the recruitment of a generation of students from families previously excluded from higher education, and so on. Nonetheless, certain characteristics were strongly associated with the idea of 'Cambridge English', even if one might also identify several other sources of their impact within English studies, and I shall conclude by briefly mentioning the three most salient.

The first was the idea that the central, defining activity of literary study was close verbal criticism. A comparison between a representative sample of books and articles written about the major figures and works of English literature in, say, the 1880s with a similar sample written in, say, the 1950s would make this point very forcibly. The texture of writing about past literature, at least that writing produced by the institutionally sanctioned 'experts' in the subject, was by the middle of the twentieth century radically different from that of earlier periods. There were other kinds of work that still retained their scholarly respectability and standing – textual editing, closely researched biographies, studies of sources and so on – but the prestige involved in 'criticism', criticism conceived as this rigorous, attentive, responsive close reading, was what gave the discipline of English its distinctive identity and, for a while, its great cultural authority.

The second characteristic associated with 'Cambridge English' was the insistence that the skills of literary analysis and judgement should be applied to recent and contemporary literature, and not just to the well-established canon of past works. A revealing detail of the English Tripos was, and still is, the fact that the most recent period paper does not have a terminal date, as it usually did in most other universities (indeed, at Oxford until past the middle of the twentieth century the course stopped at 1832). At Cambridge, the rubric of the relevant paper was always 'literature from . . . to the present'. And this concern was not a merely scholarly or pedagogical one (how could it be?). The task of literary criticism – a task to which students as well as teachers were called – required the development and application of standards of judgement, honed by exposure to the indisputable literary masterpieces of the past, to the discrimination of works of real literary merit in the present from the mass of mediocre or meretricious books promoted by the commercial needs of publishing firms and the mutual back-slapping of a coterie of corrupt reviewers.

And the third characteristic was the belief that one's response to a work of literature involved one's whole being, so that criticism was fundamentally a moral activity. Other disciplines might require a highly developed technical skill or a degree of professional judgement, but only English, it was sometimes claimed, engaged one's deepest human commitments and capacities. The adequacy (or otherwise) of one's interpretation of a particular poem or novel was ultimately an expression of the adequacy (or otherwise) of one's whole personality: shallow and limited readings were an index of shallow and limited personalities. For this reason, 'Cambridge English' certainly tended to raise the temperature in scholarly and institutional disputes: after all, so much was at stake. Outsiders might be forgiven for occasionally thinking that the chief 'Cambridge contribution' here was a marked increase in name-calling.

But above all, Cambridge English, at its best, gave those who taught and studied under its inspiration the sense that they were engaging with some of the most powerful and moving creations of the human spirit, that they were intensifying their own capacity for experience, and that they were attempting to combat the slackness of attention and coarseness of response encouraged by the commercialised environment around them. In short, it gave them the sense that what they were doing *mattered*. And that, in the highly professionalised and career-oriented world of the modern university, is a contribution beyond price.

CHAPTER 4

The Cambridge contribution to economics

GEOFFREY HARCOURT

I am a Fellow of Jesus so I must start with the person Keynes called 'the first of the Cambridge economists', Thomas Robert Malthus, as you would say, but according to Keynes, as the name is an adaptation of Malt house, the correct pronunciation is Malt-house. You may see his portrait in the dining hall of Jesus College. Keynes called him 'the first of the Cambridge economists' because he was the first chap to think like Keynes (Keynes never did consider modesty a virtue). I have a great affection for Malthus, partly because he had a stock (or perhaps a flow) of one-liners which I enjoy. In the first edition of his famous essay on population you will find some really funny remarks about the nature of the passion between the sexes which he thought was as near to a constant as would be likely to be found amongst human beings. When he was arguing with his dad, who took a Godwin stance on the possibility of perfection of humanity, Malthus, as befits a member of the Church of England, was more gloomy. He said there are two great constants: one, the passion between the sexes; the other, the fact that, as population grew, since there was a limit to the quantity of land and also to its quality, food and other necessaries would not grow as fast and so we would always be near to the constraints of starvation and misery. In fact, if we temporarily overcame the constraints, the best we ultimately could hope for was a larger population which would be just as miserable as the smaller one. So you can see why economics used to be called the 'dismal science' – we predicted maximum human misery if Malthus was correct.

I must impose two constraints on myself: the first is that there is an inbuilt tendency for sermon-givers in Jesus College Chapel to talk for thirteen minutes (not always abided by), for dissenters to talk for twenty-five minutes and for University lecturers to talk for fifty minutes; that is the first constraint. The second is that I am not going to talk about any economists who are still teaching in Cambridge, not only because I might be indulging in slander, but also because it would be invidious to try to pick out the stars and non-stars amongst my colleagues. I will say this, that they continue the Cambridge tradition in at least one aspect – we still have as many brawls as ever we had in our Faculty since we began. We have always had a propensity to produce bantam cocks who fight with great gusto on a dunghill for their positions. Of course, we do have some tremendous stars, but I am not going to tell you who they are – you have to be retired before you get a mention today. With these provisos, let me get going.

I now jump from Malthus at the beginning of the nineteenth century to the person who is responsible for the foundation of the Economics Tripos, and also for the traditional approach to economics in Cambridge, as we know them today. I mean, of course, Alfred Marshall (1842–1924), whose Library stands opposite the Austin Robinson building.

The tradition that Marshall started was the idea that economics should explain how the world worked and then do something about it, if it did not work well. This should be done by both theorising and doing applied work and then formulating feasible policies. Marshall, who was a strange and convoluted character, rather wavered on aspects of this. But the tradition was taken up, often without any inhibitions, thank goodness, by many of his pupils and followers.[1]

Marshall was not, in many ways, an admirable person, but he

was a great economist. He read mathematics at St John's College towards the middle of the last century. He was then a progressive and idealistic person, greatly influenced by Henry Sidgwick (whose lasting monument is Newnham College), as well by Sidgwick's books and contributions to philosophy. Marshall belonged to the 'Grote Club'; he went from mathematics to flirt with psychology and then into economics. He was in favour of women being allowed into Cambridge and of doing things for the poor and underprivileged. As to the latter, he did continue to think about them and contribute ideas on poverty and its causes until his death, but once he married (in 1876) an absolutely marvellous person, Mary Paley (Marshall), who had been his student, he switched sides on the women question, opposing women's entry to the University, and greatly upsetting his former colleague and supporter, Sidgwick. He told his wife that women could not do economic theory but that she could act, in effect, as his research assistant. He did allow her to be a lecturer, first at Bristol where they went after they married, then at Oxford and then finally here at Cambridge. She served him loyally, in I think a Quixotic way, and indeed she wrote a book with him, which he later suppressed even though it is one of his best books. So I do not like Marshall and I agree with my mentor Joan Robinson, who said, 'the more I learn about economics the more I admire Marshall's intellect and the less I like his character' (1973, 259).

Marshall set about taking economics out of the Moral Sciences Tripos, where it had a niche but did not, on the whole, attract good students, and making it a separate Tripos which, in the event, started in the early years of this century. It is on that foundation that we have built ever since, and though, of course, the structure has changed as the subject has changed, we owe much to Marshall for the way in which economics is taught and practised in Cambridge today. His own major contribution was his huge *Principles of Economics*, first published in 1890 and which for the first five editions was called volume 1. Marshall intended to write three (or even four) volumes but ill health (which he rather enjoyed) and certain character deficiencies meant he never was able to write down in a systematic form the other volumes. In old age he published two more volumes which were not a patch on the famous first volume. He died in his early eighties without being able to complete his project. He did establish what is known as 'The Cambridge Oral Tradition', that is, Cambridge chaps would say to the lesser breeds without the Law: 'oh well, of course, we always talk about that at Cambridge – it is in the Oral Tradition, it may not be written down, but that is what we mean' – a good ploy to fall back on, especially when on dangerous ground. People obtained their notion of what Cambridge economic principles were about, partly from the *Principles*, partly from articles and memoranda to Royal Commissions and so on; and partly from people's views and recollections of his lectures and supervisions (he was an excellent supervisor).

What was the structure of Marshallian economics? Putting it simply and not doing full justice to its richness, in the first volume he wrote about the nitty-gritty of economic life – what determines the prices and quantities of commodities, what determines the employment, wages, and salaries of different classes of labour, what determines the rate of interest and the rate of profit in various industries: what we now call the theory of relative prices and quantities. His first great analytical contribution in this endeavour was to introduce systematically into economics the use of supply and demand functions and curves. You have often heard it said that economists are just parrots who say 'supply and demand, supply and demand'; if anyone is responsible for us being parrots, it is Marshall. But he was no parrot; rather, he used the supply and demand apparatus to handle in a systematic and rigorous manner the analysis of the determination of prices and quantities in mainly competitive markets. His second contribution was to recognise in a deep way that the most difficult and yet relevant concept which affects economic life is time. You know the saying, 'Time is a device to stop everything happening at once', which is said to be due to the philosopher, Bergson. A group of us in the

1960s were discussing who actually said this. An Indian economist, Dharma Kumar, a Cambridge graduate on leave at the time, said, 'I don't know who said it, but I think that space is a device to stop everything happening in Cambridge' – one of the best spontaneous one-liners I have ever heard. A lot of people would have killed to have said it and indeed many people, though they have not killed, have certainly plagiarised Dharma ever since. In order to handle this intractable concept of time Marshall used three analytical concepts: the market period, the short period, and the long period. The market period is an immediate one; in the central market in Cambridge there will be a length of analytical time during which prices will be set such that the *given* quantities will be bought. The short period is an analytical device which refers to a period that is long enough for the number of people employed in a firm or industry to be changed, in order to change their rates of production, but not long enough for firms to either leave or enter an industry or bring in new capital goods to increase their capacity. The long period is a period of time long enough to allow both the supplies of skilled labour and of capital goods in industries to be changed, and for new firms to enter and old firms to exit. Marshall made clear that these are not one-for-one descriptions of real life, but analytical devices which use the concept of *ceteris paribus* (other things being equal); he called it the *ceteris paribus* pound, in order to allow us to get a grip on what otherwise is an intricate interconnecting process which is impossible to make sense of in a systematic way. In volume 1 he used these devices in order to go systematically through the determination in the market period, the short period, and the long period, of prices and quantities of commodities and prices and employment of the services of the factors of production; and so to develop theories of rent, wages, profits, and interest – all in the supply and demand framework. Strangely enough though, because he was a most realistic person, money did not get a mention except as a ticket – as something with which to measure things. Everything was done in real and relative terms and money might just as well never have existed as far as analysis was concerned. This is a bit unfair for we now know, as a result of the recent publication of the first major biography of Marshall, by Peter Groenewegen (1995), that Marshall was more guarded than this. He always qualified and modified, and whenever you thought you had something new, it would be said, 'there's nothing new about that, it's all in Marshall'. But it was the message which people took away from volume I. In volume II, if he had ever got round to writing it down systematically, he meant to talk much more about money and monetary institutions and, combined with this, he was to talk about what determined not the prices of individual commodities or of the individual services of factors of production, but the general price level – the concept of the price level of all goods and services in aggregate in the economy. In doing that he would have been one of the first to develop in a systematic way the theory of the general price level which we call the quantity theory of money, which has made a great comeback in recent years via the monetarists and in particular, their high priest Milton Friedman of Chicago, a Marshall admirer.

Marshall developed the quantity theory of money in order to try and describe what determined the general price level; he argued that, at least in the long period, what was happening in the real sector of the economy concerning employment, production, and relative prices, and what was happening in the monetary sector of the economy, the banks, the financial sector generally and the formation of the general price level, were independent of one another. Money was basically a veil. We all know what may be done with a veil; as we are getting to mature years, we could pull it down and provide an aura of mystery; if we are younger we could lift it up. But the vital point is, that what ever is underneath is not affected by the veil itself. That is how, generally speaking, money was treated in the Marshallian tradition – at least in the long period. It is true that in the short period, when looking at the workings of the economy as a whole, it was admitted that money could have real effects on the economy but this was not worked out systematically or satisfactorily because they were constrained by the dichotomy between the real and the monetary.

Marshall, of course, contributed many, many things besides those I have alluded to,[2] but the important thing that I want to emphasise here in the context of the Cambridge contribution to economics is that he systematically developed the idea of the real sector and the monetary sector and the quantity theory of money as an explanation of the general price level in the long term. This meant that the role of the monetary institutions, including central banks, was to make sure that they so controlled the monetary side of the economy that the underlying real things operating in a competitive environment would not be handicapped in their determination of the allocation of resources, with supplies and demands responding to each other. The basic idea was that if we had a lot of competition, on the whole the level of activity and the composition of goods and services that were produced would be responding to what people wanted, as expressed through their demands and reflected in the price mechanism.

It would be wrong to say that Marshall and his followers were uncritical defenders of *laissez-faire*, that if you leave it alone, the system will work well. They were not. They recognised that the system had deficiencies, that there was poverty, unsatisfactory working conditions, and lapses from full employment, and that there *was* a role for government intervention. Yet, on the whole, they were great supporters of creating competitive institutions and then letting the price mechanism do its thing. This was the underlying philosophy to which they admitted definite exceptions, the extent of which varied according to their particular philosophies, personalities, and so on. Nevertheless, they argued that there were strong forces, if there was competition, which would not only ensure that the goods and services produced were the goods and services that people wanted, but also that there was a tendency for people who wanted to work and capitalists who wanted to employ their capital in particular ways, to be able to do so. And, logically, that is what they had to believe, because if there

GEOFFREY HARCOURT

were not a tendency to full employment, it would not be possible
to argue that, even for the long period, the quantity theory was an
explanation of the general price level because (for those of you
who are mathematicians) there would be three unknowns with
only one equation. Whereas if it were argued that there was a
tendency for prices, that is real wages, in the labour market – just
as in any other market – to settle at the point where what was
voluntarily supplied as labour services was voluntarily demanded
by employers in aggregate, we could find what the long-period
level of activity would be. We could then feed the answer into the
quantity theory equation, make an argument about how the
velocity of circulation of the typical pound was determined by
custom and history, and assume that the monetary authorities
controlled the quantity of money, so that the only thing not
known would be the general price level. With one equation, and
one unknown, we could solve the general price level. That was
the thrust of Marshall's teaching, and of the Cambridge teaching
of monetary theory generally, up until the end of the 1920s.

I now pass on from Marshall. He retired in the early years of this
century, to be succeeded by his protégé, A. C. Pigou (1877–1959),
who was in his early thirties. (Those were the days of child
professors.) Pigou carried on Marshall's work, though not always in
ways Marshall liked because Pigou was much more of a bolshie
than Marshall turned out to be. He drew on Marshall's work about
how the price mechanism did not always do its thing correctly, and
wrote a book which went into several editions, *The Economics of
Welfare* (1920). Pigou's views still influence us today. He pointed
out that, often, the social costs and the social benefits of production
do not match their private counterparts. Because business people
are guided by their private gain (why should they be guided by
anything else, they are not there as altruists, they are in business to
make profits – that is what capitalism is about), we may often get an
allocation of resources, a level and a composition of production,
which do not take properly into account their social costs and
benefits, which in our jargon are called 'externalities'. The best-

72

known example is that of a factory which belches out smoke, yet its owners do not have to pay the costs of people living nearby who have to do extra washing – yet they ought to. These ideas are now applied to how we may handle the problem of destroying our environment by taking into account the social as well as the private costs of production. Piero Sraffa, who was the Marshall Librarian for many years, had Marshall's copy of an earlier version of Pigou's book, in which Pigou made these suggestions for government intervention; it was full of Marshallian annotations saying things like: 'Oh, he shouldn't have said that, he shouldn't have said that at all.' He wondered what he had let loose on the world, this bolshie chap, Pigou, who was going much farther than he himself was prepared to go. *So far* that in the ridiculous hunt for spies after the Second World War (you know, the Cambridge spies and all that), Pigou was mistaken for a Russian spy – absolute piffle. So, Pigou developed arguments about the sort of government interference which related to social costs and social benefits. He had a most illustrious and important career. Now he is a rather forgotten figure, but he should not be forgotten at all.

But, of course, Marshall's most distinguished pupil was John Maynard Keynes. He dominated Cambridge economics from the 1920s to his death in 1946 – and beyond. Keynes is one of my great heroes. For though he sometimes seemed to have had feet of clay, my judgement is that in a fundamental sense, he was a *good* man. I interpret his life as trying to solve the conundrum posed by the Cambridge philosopher G. E. Moore (who was a defining influence on the young Keynes and other bright young things around Cambridge at the turn of the century): is it possible both to *do* good and *be* good? I think Keynes's life provides a resounding *yes* to that question. His Bloomsbury friends were less sure. They, or some of them anyway, rather drew apart from the world and lived individual lives – we know the sort of lives they lived, they had endless affairs to make copy for their next novels and so on – a circle who loved in triangles and lived in squares, as Lord Annan recently reminded us. Keynes was a naughty man in lots of ways but he was driven, as were/are all the outstanding Cambridge economists, by an intense seriousness: a desire to understand the world and then make it a better place. Keynes was a civil servant as well as a don, and a very courageous man once he grew up. He started off as Marshall's pupil, but in the end he was to overthrow, in fundamental ways, the Marshallian legacy; not out of lack of piety for his teacher but because he followed arguments wherever they led, no matter how unpalatable their conclusions were. He was also very much a man of affairs.

It is interesting that virtually all these people come from either the middle or upper middle classes of British society except for poor old Marshall, who was barely lower middle class. When he died his wife connived with Keynes, consciously or unconsciously, to hide what his origins were. For example, he had given his father a more posh role in banking circles than he actually had and he hid the fact that his mother was the daughter of a butcher and the granddaughter of an agricultural labourer (see Coase, 1984, 520); which was silly because Marshall's view of economic progress allowed people in our sorts of society to realise their potential. Indeed he was an excellent example of someone who came from rather lowly origins yet realised his own potential as a professor at Cambridge, but he had that sort of inverted snobbery which made him suppress his origins, whereas his wife was much higher up the social pecking order and so she did not worry about such things; the same was true of Keynes, Austin and Joan Robinson. It gave them a confidence, even an arrogance and imperiousness, which allowed them to think that they could actually do things, that they could fulfil what Harrod (1951) said of Keynes, 'the pre-suppositions of Harvey Road'. (Harvey Road was where the Keynes family lived.) The idea was that there was this group of disinterested people who worked in the civil service, the universities, and the public schools, training intelligent people to go out and find out how the world worked and make it work better; in particular, to stop the malfunctioning of society falling on those who are least able to defend themselves. A noble idea, I think.

Keynes started as a mathematician but he was only twelfth wrangler, a respectable result but not what was expected of him. He therefore read for the civil service exams, sitting at Marshall's feet. Marshall quickly realised that he was a brilliant student. He said, in effect, we old men will have to kill ourselves, the world's only safe for the young now. Keynes went into the civil service, getting his worst marks, by the way, in economics. 'Presumably, the examiners knew less about the subject than I [Keynes] did'. He was elected to a Fellowship at King's College in 1909, and came to back to teach at Cambridge before the First World War. In the war he went into the Treasury; he was at the Treaty of Versailles as a junior assistant to Lloyd George. Keynes was so horrified at what the French (and the Australians, through Billy Hughes) and Lloyd George were doing to Woodrow Wilson and through Wilson, to the Germans, that in the end he resigned and wrote the book which first made him famous, *The Economic Consequences of the Peace* (1919). In it he examined how the pre-war European economy worked and how the vicious reparations that were to be imposed on the Germans would not only wreck Germany's economy but would also disturb the delicate balance of how things were done in Europe, and bring about a catastrophe as well as being inhumane and ungenerous to a defeated enemy. It brought him fame but it also put him in the wilderness as far as official circles were concerned for the inter-war years (though not as much as was first thought).

In writing the book he was still applying Marshallian principles; it was during the 1920s and especially in the 1930s that he started to rethink drastically about how the world worked. Of course, he was not alone; in the 1920s his closest ally was Dennis Robertson (1890–1963). They had a most productive intellectual partnership and friendship which, alas, did not survive the making of *The General Theory*. This was both a personal tragedy and a professional one for the development of economic theory.

As Keynes was rethinking Marshall's monetary theory, others at Cambridge were starting to rethink Marshall's theory of the determination of prices at the firm and industry level. The latter development was especially associated with four people. The first was an Italian émigré chased out of Italy by Mussolini, Piero Sraffa, who was one of the most important intellectual influences of the twentieth century. He was the intimate friend of Keynes, Wittgenstein, and Gramsci. He published a most important article in the *Economic Journal* in 1926 (Keynes was co-editor with Edgeworth); it was principally an attack on Marshall's method of doing economics. But Sraffa suggested as well that rather than having competition as the general model of how markets worked and industries and firms behaved, monopoly – the other end of the spectrum – would be more appropriate. Sraffa said that if you asked business men, 'Why don't you sell more? Is it because your costs are rising?', they would laugh you out of court, adding, 'We can't sell more because we'd have to cut our prices too much, so it wouldn't be profitable to do so.' Sraffa suggested that we ought to look at the formation of prices and quantities in modern industries as resulting from mini-monopolies surrounded by other mini-monopolies, so that they had to take account of their customers' reactions and other firms' reactions when they set their prices. This was a huge blow to the case for *laissez-faire* because one of the arguments for competition is that it rids the system of the unfit. Marshall (and Pigou, following him), predicted that if there was a fall in demand for products, unfit firms would disappear, only the fittest would survive. (Marshall was not alone in this but we are talking about the Cambridge tradition.) Sraffa showed that this was not true, that most firms would survive but they would be working at under-capacity and therefore competition was not the effective *clean it out* sort of process that it was thought to be.

Sraffa's article precipitated what became known as the imperfect competition revolution. It was developed by Gerald Shove, Richard Kahn (1929, 1989) (Keynes's favourite pupil, in King's), and then by Austin and, especially, Joan Robinson, who published *The Economics of Imperfect Competition* in 1933, in which these ideas were synthesised and systematically expounded. At the same time as the revolution in price theory was occurring here it was independently occurring at the other Cambridge. (For example, Edward Chamberlin of Harvard published *The Theory of Monopolistic Competition* in 1933.)

Keynes himself became more and more dissatisfied with Marshall's way of looking at the workings of the economy as a whole. In particular, he discarded the argument that we could talk about prices and quantities and employment *independently* of what was happening in the financial sector and in the monetary sector generally. He had a go at solving this in what was meant to be his *magnum opus*, *A Treatise on Money*, which was published in two volumes in 1930. In many ways it was a continuation of the Marshallian tradition. But he was changing his emphasis from the long period, the central core of Marshall's economics, to the short period, though he continued to regard the latter as stations on the way to the long-period cross. In fact, as early as 1923, cheeking Marshall, he had written: '*In the long run* we are all dead. Economists set themselves too easy, too useless a task if in tempestuous seasons they can only tell us that when the storm is long past the ocean is flat again' (1971, 65, italics in original). He tried increasingly to design monetary policy which contained measures which would stop or at least ameliorate the effects of the huge inflations and deflations that had occurred after the end of the First World War. This runs through his work in the 1920s, but it was never satisfactorily done.

So in 1930–1 he started again. In this task he was aided by a remarkable group of young economists ranging from about twenty-six to thirty-three in age who were here in Cambridge and who came together in what was called the Cambridge 'Circus'. The principal members were Austin and Joan Robinson, Piero Sraffa, Richard Kahn, and James Meade (who died just before Christmas 1995, the last 'Circus' person to die). Meade had come from Oxford to study economics before he took up a Fellowship at Hertford College. Meade spent a year here and they discussed Keynes's *Treatise on Money* and they helped him (but he, of course, was the major author) to develop what became his authentic *magnum opus, The General Theory of Employment, Interest and Money* (1936).

What did he do in it? As you remember, the 1920s in Britain and then the 1930s all round the world were characterised by terrible mass unemployment, which in those days was thought to be a sin. (These days it is thought to be necessary in order to provide a quiescent and cowed workforce to allow international capital to prosper. Marx understood this much more, I think, than Marshall.) In those days people were shocked by mass unemployment; they were trying to work out why it had occurred because economic theory, by and large, said that, at least in the long term, it could not occur if the impediments to competition working were removed. What Keynes finally decided when he worked through the traditional analysis again was that it was wrong. He went back to his hero Malthus, who had had an argument with another great economist, David Ricardo, about whether it was possible to have a general glut, that is, could there be a failure of *overall* demand? (Everybody accepted that there could be gluts or scarcities in individual markets because individual demands and supplies had not matched temporarily.) But could there be a failure of demand *overall*, so that people and machines lay idle? Ricardo's answer, based on the argument of the French economist, Jean Baptiste Say, was, 'No, commodities buy commodities, supply creates its own demand.' There cannot be a lack of aggregate demand in the long term. With competition everything will match up, including the labour market where the real wage rate will settle at a level where those who want to work are able to find jobs. Basically that was the argument that Keynes attributed to Ricardo (though now we would say that, while it was implicit in Marshall, Ricardo only argued that machines, i.e. capital, could not be idle in the long term). Malthus countered that there could be a deficiency of overall demand, but he never could explain satisfactorily to Ricardo why this was so. Keynes argued (he was not a good historian of thought) that Ricardo conquered Malthus and that from then on we had all believed in Say's Law. We had a theory which deduced that there could not be a general glut, at least as a long-period proposition, in a world where there was up to 29 to 35 per cent unemployment. The theory was nonsense, the practice just did not match. Keynes set out to provide a theory that did. In doing that he argued, following Malthus, that as well as there being aggregate supply we had to have a theory of aggregate demand, a theory which determined the total demand for the production of the economy and therefore the total employment of people in it. He argued that one of the important components of aggregate demand was investment expenditure or capital accumulation – new factories, machines, and so on. A crucially important determinant of investment was the uncertain future – we do not know what is going to happen in the future but in order to invest we have to make guesses about what is going to happen, what we think our future profits will be, and then invest.

Keynes showed that basically there were not persistent forces at work in the economy which, at least on average, would produce enough investment expenditure at any moment of time to absorb the amount which people in the economy would be willing to save voluntarily if they were receiving the incomes they would get if the economy was at full employment: that is to say, if everyone who wanted to work could find a job at ruling wage rates.

What Keynes claimed to have demonstrated was that for prolonged periods of time the economy could settle at much lower levels than full employment, where what people voluntarily saved equalled what people were willing to invest but which left many people and machines idle. The unemployed were willing to work but there was no way of signalling to the people who were trying to make profits, that it would be profitable to employ them. And, indeed, it would not be unless there were to be a rise in aggregate demand as a result. It followed, therefore, that there was a case for government intervention.

One reason why economists had not seen this clearly before was the dichotomy between the real and the money – with money being just a veil. Whereas Keynes said, 'well, let's think about it again. Money is not a veil. Money is there right from the start.' Why? Because one of its properties is that it is a store of value, as well as a medium of exchange. If people are uncertain about the future they can hold money rather than spend it. Therefore, there is a second reason for holding money which plays an important part in determining what the pattern of rates of interest on financial assets is. He worked out a theory of what these patterns of rates of interest were, and he argued that business people have a fundamental choice – either they can hold financial assets on which they get interest payments, or they can do real things like invest. He argued that often rates of interest settled at levels where the amount of investment that people would make before the expected returns to investment had fallen to the level of the rate of interest (the alternative thing to do) was such that aggregate investment was not large enough to absorb full employment saving. Therefore, if we had a theory of the economy where money played a role right from the start and we did not have a dichotomy but an integrated theory, we could show that there were tendencies to unemployment – to a failure of aggregate demand.

That was his great contribution[3] and the Cambridge contribution to economics in this century has built on this foundation ever since: first, in developing policies to run a wartime economy, including keeping inflation in check; secondly, in the post-war period in developing longer-term theories of growth and distribution over time. Here, some of Keynes's younger colleagues from the 'Circus', together with Nicholas Kaldor (who came to Cambridge from the London School of Economics after the war), Richard Goodwin and Luigi Pasinetti, came into their own in developing the peculiarly Cambridge contribution to theories of growth and distribution, the issues which had been the preoccupations of the great classical economists, Adam Smith, Ricardo, and Malthus, leading on to Marx. That was one aspect of what came out of Keynes's and his colleagues' contributions.

Another aspect was the development of ideas of which Marshall would have approved, though he did not do much himself; this was the often down to earth, pioneering applied work associated with the Cambridge Department of Applied Economics. The first Director was a marvellous man called Dick Stone (Sir Richard Stone), who died a few years ago and who, with James Meade, is the only overwhelmingly Cambridge economist to get the Nobel Prize. (Now we may add Jim Mirrlees, if I may break my vow not to mention any colleagues, and also overlook his twenty-five years or so at Oxford.) Starting in 1945, Stone headed up an extraordinary group of researchers and research in the Department of Applied Economics (DAE). One of the things for which he was known (and for which he received the Nobel Prize) was his development of National Accounting, which is a way of looking at the production, expenditure, and income that is produced annually in a Keynesian sense. The structure that Stone developed reflects the theoretical developments of Keynes in his theory of aggregate demand and aggregate supply. (Meade and Stone pioneered the structure of such accounts early in the Second World War. Austin Robinson recruited them for this task and he regarded their achievements as his greatest contribution to the war effort.)

Dick Stone not only presided over developing world-wide national accounting standards, he also was a pioneer of demand theory and practice. Here Marshall's contributions were influential in a conceptual sense while Stone was one of the pioneers of the actual *estimation*, through econometric and statistical methods, of demand curves for various commodities. Stone presided over a remarkable ten years in which what are now some of the most famous names in our subject produced highly original econometric and applied work. I should especially mention the seminal work on economic history, using a Keynesian framework, associated with Max Cole, Phyllis Deane, Charles Feinstein, Robin Matthews, and Brian Mitchell, which commenced under Stone's benign leadership and encouragement. In a sense this filled an essential gap associated with Marshall's inability to complete his original project.

The tradition which Stone started has been continued, with different emphases, by the various Directors. Brian Reddaway took over from Stone in 1955 (Stone became the P. D. Leake Professor of Accounting and led with J. A. C. Brown the growth project in the DAE). Reddaway undertook himself (and encouraged many others to do likewise) applied projects characterised by the 'Reddaway Method' – thorough knowledge of data and its limitations, careful statistical analysis of it and of what it can be used to show, and what it cannot. When Brian became the Professor of Political Economy in 1970, Wynne Godley, as Director (1970 – 87), carried on both the Keynes and the Marshall stories, because he was very interested, first of all, in the role of forecasting in economic policy and, secondly, at a theoretical level, in showing how in the short term and in the long term the real and the monetary aspects of the economy may be combined together in a consistent set of stock and flow accounts. He was then succeeded in 1988 by David Newbury, about whom I may only say that I used to beat him at squash (I think!). So, alongside the theoretical developments I discussed earlier, we also have a tradition of applied work.

As I noted above, in the post-war period some of those from the 'Circus' were joined by the man who most resembled Keynes in the post-war period – Nicky Kaldor, who was a larger-than-life figure: a completely honest and ultimately lovable man who always said what he thought, who loved and lived life to the full. Kaldor, Joan Robinson, Richard Kahn, Piero Sraffa, Richard Goodwin, and Luigi Pasinetti between them developed theories of growth and distribution peculiarly associated with Cambridge, and also, in my view, a damning critique of the reversion to pre-Keynesian theory associated with the monetarists, and of the theory of value and distribution which was associated with Marshall amongst others. They felt that there were serious flaws in the latter's conceptual foundations. They criticised it in a number of important books and articles while at the same time developing alternative approaches which not only drew on Keynes's insights, but also reached back (not in a retrogressive but in an historical sense) to the classical political economists. They adopted their central concept of the surplus, its creation, extraction, distribution, and use, to give us a new way of thinking about the economy. Included in both these endeavours we may mention, first, Sraffa's edition of the works and correspondence of Ricardo, on which he worked from the early 1930s and which, in collaboration with Maurice Dobb,[4] came to fruition in eleven volumes in the 1950s (1951–73). (The index was published in the 1970s.) Sraffa's Introduction to volume 1 and his 1960 classic, *Production of Commodities by Means of Commodities* laid the foundations both for the critique of the supply-and-demand theories of mainstream economics and the revival of the classical Marxist method and approach through the concept of the surplus, which was incorporated into the peculiarly Cambridge approach to value, distribution, and growth theory. Secondly, we may name Joan Robinson's *magnum opus, The Accumulation of Capital* (1956), her many influential works and articles that cluster around it, and a whole host of Kaldor's articles starting in the 1950s. His ideas reached their final form in his Raffaele Mattioli Lectures. They were given in May 1984, just two years before he died, but have only recently been published as *Causes of Growth and Stagnation in the World Economy* (1996). In this endeavour they were joined by the Polish Marxist economist, Michal Kalecki, who independently discovered the main propositions of Keynes's *General Theory*. As he came from a Marxist background, he found it natural and easy to link his findings to classical political economy as fulfilled in Marx's analysis of capitalism. (While I think Marx on 'how to run an ideal society' can be even more Utopian than the most Utopian Christian Socialist, as an analyst of capitalism he, Kalecki, and Keynes had no peers – Marx understood capitalism better than probably anyone else.) Kalecki, who belonged to that tradition, very much influenced Joan Robinson's writings.

Richard Goodwin (who died in August 1996) was a most original and eclectic economist. He absorbed all the elements set out above, together with those that came from his mentors, Joseph Schumpeter and Wassily Leontief at Harvard, to produce a theory of cyclical growth, the indissolubility of trend and cycle. It emphasised both aggregative trade-cycle theory and the production interdependent systems associated with Sraffa and Leontief (see Goodwin and Punzo (1987)). Luigi Pasinetti (Goodwin's pupil at Cambridge) has over a thirty-year or more period developed a unified system of distribution and growth which absorbs both classical and Keynesian ideas. He is perhaps the last great system builder of our profession (see Pasinetti 1981, 1993). Finally, on the subject of growth theory, the profession will ever be indebted to Frank Hahn and Robin Matthews for their masterly survey article of the state of the art in growth theory. It was published in the *Economic Journal* in 1964, and it set the standard for survey articles from then on.

I must also add that Marshall's original concern with poverty and injustice and their cures lived on in many of Meade's writings and in the writings of Tony Atkinson (who was inspired by Meade and about whom I may talk as he has gone to Oxford) on the distribution of income and wealth and the causes and cures of poverty.

Well, that is what I think the Cambridge contribution and tradition are about. I am a fortunate person because I have been a student in this tradition, I have taught it, most of the people about whom I have written (with the exception of Keynes, and of course Pigou, Marshall, and Marx – I am not *that* old) were my teachers, and then my colleagues and my friends. So, it has been my good fortune also to work in this tradition. I have brought to this peculiarly Australian contributions as well, because of the mentors I had in Australia, most of whom came out of the Cambridge tradition. Therefore, you see before you a fulfilled person who has tried to do something to preserve the Cambridge tradition and who is going to play cricket for the Jesus College Long Vacation

High Table side this afternoon. On a glorious summer day, what more could anybody ask for?

This chapter is based on the transcript of the lecture I gave in the Lady Mitchell Hall to the Summer School in July 1996. I have tried to make the text grammatical but otherwise I have left it much as it was delivered.

1 In writing on the Cambridge approach to applied economics, Michael Kitson and I (1993, 437) put it as follows: 'it emphasises the importance of relevance in economics, incorporating the lessons of history, the international context and prevailing social and political conditions. Theory and measurement are mutually interdependent as robust empirical analysis is dependent on relevant theory, which in turn depends on reliable observations. Cambridge advances in theoretical and applied economics have . . . gone hand-in-hand . . . techniques have never been allowed to obscure the analysis – the medium is not the message.'

2 Not least his method of partial equilibrium analysis – looking at particular parts of the economy, locking the rest in the *cet.par.* pound, in order to be able to say something concrete. He understood about general equilibrium analysis but thought it could not get much past the profound insight that everything depends upon everything else.

3 It must be added that Richard Kahn was especially influential in the making of *The General Theory*: first, as a remorseless critic of the quantity theory as a causal explanation of the general price level; secondly, through his own remarkable work in the late 1920s on the economics of the short period (Kahn 1929, 1989) in which he made the short period a subject worthy of analysis in its own right, thus reinforcing Keynes's own inclinations; and, thirdly (with James Meade), through his 1931 article on the multiplier which provided the central, indeed, crucial concept for Keynes's new system.

4 For many decades Maurice Dobb was the foremost Marxist economist and scholar in the United Kingdom (he died in 1976). He wrote

several classics which remain sources of inspiration and instruction. My favourites are *Political Economy and Capitalism* (1937), *Welfare Economics and the Economics of Socialism* (1969) and *Theories of Value and Distribution since Adam Smith* (1973).

REFERENCES

Chamberlin, E. H. 1993. *The Theory of Monopolistic Competition. A Reorientation of the Theory of Value* (Cambridge, Mass.: Harvard University Press)

Coase, Ronald H. 1984. 'Alfred Marshall's mother and father', *History of Political Economy*, 16, 519–27

Dobb, Maurice. 1937. *Political Economy and Capitalism. Some Essays in Economic Tradition* (London: Routledge)

   1969. *Welfare Economics and the Economics of Socialism. Towards a Common-sense Critique* (Cambridge: Cambridge University Press)

   1973. *Theories of Value and Distribution since Adam Smith. Ideology and Economic Theory* (Cambridge: Cambridge University Press)

Goodwin, R. M. and L. F. Punzo. 1987. *The Dynamics of a Capitalist Economy* (Oxford: Polity Press)

Groenewegen, Peter. 1995. *A Soaring Eagle: Alfred Marshall, 1842–1924* (Aldershot: Edward Elgar)

Hahn, F. H. and R. C. O. Matthews. 1964. 'The theory of economic growth: a survey', *Economic Journal*, 74, 779–902

Harcourt, G. C. and Michael Kitson. 1993. 'Fifty years of measurement: a Cambridge view', *Review of Income and Wealth*, Series 39, No. 4, December, 435–47

Harrod, R. F. 1951. *The Life of John Maynard Keynes* (London: Macmillan)

Kahn, R. F. [1929] 1989. *The Economics of the Short Period* (London: Macmillan)

   1931. 'The relation of home investment to unemployment' *Economic Journal*, 41, 173–98

Kaldor, Nicholas. 1996. *Causes of Growth and Stagnation in the World Economy* (Cambridge: Cambridge University Press)

Economics

Keynes, J. M. [1919] 1971. *The Economic Consequences of the Peace* (London: Macmillan, *Collected Writings*, vol. II)

   [1923] 1971. *A Tract on Monetary Reform* (London: Macmillan, *Collected Writings*, vol. IV)

   [1930] 1971. *A Treatise on Money*, 2 vols. (London: Macmillan, *Collected Writings*, vols. V, VI)

   [1936] 1973. *The General Theory of Employment, Interest and Money* (London: Macmillan, *Collected Writings*, vol. VII)

Marshall, A. 1890. *Principles of Economics* (1967, Variorum edn) (London: Macmillan)

Pasinetti, L. L. 1981. *Structural Change and Economic Growth: A Theoretical Essay on the Dynamics of the Wealth of Nations* (Cambridge: Cambridge University Press)

   1993. *Structural Economic Dynamics: A Theory of the Economic Consequences of Human Learning* (Cambridge: Cambridge University Press)

Pigou, A. C. 1920. *The Economics of Welfare* (London: Macmillan)

Robinson, Joan. 1933. *The Economics of Imperfect Competition* (London: Macmillan)

   1956. *The Accumulation of Capital* (London: Macmillan)

   1973. *Collected Economic Papers*, vol. IV (Oxford: Basil Blackwell)

Sraffa, Piero. 1926. 'The laws of returns under competitive conditions', *Economic Journal*, 36, 535–50

   1960. *Production of Commodities by Means of Commodities: Prelude to A Critique of Economic Theory* (Cambridge: Cambridge University Press)

Sraffa, Piero with the collaboration of Maurice Dobb (eds.) 1951–73. *Works and Correspondence of David Ricardo*, 11 vols. (Cambridge: Cambridge University Press)

CHAPTER 5

'Nasty forward minxes': Cambridge and the higher education of women

GILLIAN SUTHERLAND

The theme of this collection of essays is the contribution made by Cambridge. In many fields this translates straightforwardly into an exciting story, an account of distinguished and innovative work in a particular academic discipline, or even the creation of a new one. In the field of women's education the task is not so simple. To ask what Cambridge University has done for women is to evoke a response mostly in the negative. The University resisted the initial arrival of women. When they could no longer be ignored, they were kept at arm's length: such acceptance as there was, was grudging and partial. Cambridge was the last of the old British universities to give full membership to women, waiting until 1948, after the Second World War. Even in the 1990s the proportions of women at every level are somewhat lower than elsewhere in the United Kingdom.

The question of contributions needs turning round: we should instead ask what the women contributed to Cambridge, despite the hostility and all the difficulties. The answers to this are both more positive and more complex. There are contributions within Cambridge to be explored and measured: there are contributions to the wider world of which Cambridge may be proud.

It is a large subject and this essay can serve only as a preliminary sketch. In it I shall focus particularly on the period including the two world wars and the decades between, a period which so far has received much less attention than either the earliest beginnings or the years since 1948, yet one of great importance in the evolution of the two women's colleges as institutions and in their relations with the University.

In exploring the contributions of the women in this period I shall try also to consider how they might be used to reflect on the dilemma of defining equality. For those in pursuit of equality have a choice: is it to be measured by identity, are women and men equal and seen to be equal only when they can do and do do exactly the same things? Or can there be an equality of difference? Might women make a contribution which had its distinctive features but one equally valued with that of the men – and if so, how might that equality be measured and expressed?

I

The quotation in the title was chosen to give the flavour of a great deal of the early opposition to women. 'Nasty forward minxes' was the comment of Adam Sedgwick, the distinguished geologist, in 1865 when he learned that the University had accepted a petition to allow girls' schools to enter their pupils for the examinations for secondary schools across the country devised and run by Cambridge.[1]

Perhaps Sedgwick used even stronger language when women with pretensions to be undergraduates began to arrive here at the beginning of the 1870s. In 1870 courses of lectures for women were arranged and in 1871 a house was leased in Regent Street to accommodate those wishing to attend but living too far away to travel daily. There were five of these students at first, but demand grew rapidly. By 1875 the promoters of this scheme had leased land in West Cambridge and erected the first of the buildings of Newnham College. In 1869 Emily Davies had opened her college for women also with five students at Hitchin, strategically placed, she hoped, between Cambridge and London. In 1873 this embryo institution likewise moved, to Girton on the outskirts of Cambridge, to the site which Girton College still occupies.[2]

In the years up to 1914 the women had their successes. In 1881 they gained formal permission to take University examinations, although they were not to be rewarded with degrees. Their names were to be published in a separate list, ordered by reference to the men's list. Soon afterwards, in 1887, Agnata Frances Ramsay of Girton was placed ahead of the top man in the Classical Tripos examinations. In 1890, Philippa Fawcett of Newnham was placed ahead of the top man or Senior Wrangler in the Mathematics Tripos. Between them they demonstrated the capacity of the women to compete at the highest level.

The experience of the women students also provided ammunition for those who wanted to replace the University's curious entrance examination, the 'Previous', with a wider range of qualifications, such as those increasingly being offered by good secondary schools.[3] However Sedgwick's comment, a barely concealed suggestion that women might use their sex to secure an unfair advantage, a mixture of antagonism and anxiety with a sexual edge, remained typical of attitudes in many parts of the University. There was sustained and entirely successful opposition to all proposals that women should be allowed anything other than access to University examinations, underscored by very ugly and violent demonstrations in 1897.[4]

During the First World War British women in general kept out of the front line. Few, if any, of the Girton and Newnham students emulated the young Vera Britten from Somerville College, Oxford and threw up their courses to go to train as nurses.[5] More commonly they went into war work on graduating. While undergraduates they sponsored and supported one of the units of the Scottish Women's Hospitals, which worked first in France, then in Salonica; and a number of them went to work for that heroic organisation on graduating, as doctors, as nurses, orderlies, couriers and interpreters.[6] Among those who saw service as interpreter, courier *and* orderly was Elsie Butler, who while working as a temporary lecturer in French at Newnham in 1915–16 had been furiously learning Russian in order to join the war effort. She wrote much later of:

> The release from the fearful strain of living under the cloud of war as a useless civilian ... yet everything one did was vital, urgent and exacting ... when at last I got down to Reni on the Danube with my nurses, the last great Russian offensive was in full swing, the wounded were pouring into the hospital in hundreds and even the most ignorant and inefficient were needed.[7]

Less exotic but no less important was work on the Home Front. Lucy Davis Cripps of Girton worked as a Medical Officer of Health for the Ministry of Munitions, first in Coventry and then at Head Office in London. Joan Pulling, also of Girton, became Secretary to the Editor of War Emergency Regulations and Manuals, no doubt doing a great deal of the drafting herself.[8] The lovingly hand-written record of the war work of Newnham students, completed in 1922, lists over 600 names. They did an extraordinary variety of jobs, from the medical, though war-related research and work for thirty government departments or boards to relief for refugees and prisoners of war, including a category 'substitutes for men' as teachers and administrators. The volume ends with a roll-call of the decorations earned, not only from the British Government, but also from the French, Serbian, Russian and Romanian governments.[9]

At the war's end it looked for a brief moment as if the changes wrought in attitudes and behaviour might bring major institutional and structural change.[10] The Franchise Act of 1918 granted women over thirty the parliamentary franchise. The 1919 Sex Disqualification Removal Act launched a more general removal of barriers. By the end of 1920 the University of Oxford had completed the procedures for granting its five women's colleges and their members full membership of the University.

Yet in 1921 the Cambridge men voted again to reject full membership for Girton and Newnham. Yet again the majority for rejection was celebrated with violence: using a handcart as a battering-ram, a mob of male students smashed the lower half of the beautiful bronze gates erected at the head of Newnham Walk in 1894 as a memorial to Newnham's first Principal, Anne Jemima Clough. What eventually did secure a majority at the end of 1921 was a 'half-a-loaf' proposal: that women should be granted the titles of degrees but none of the privileges that went with them.

The intransigence of Cambridge greatly embarrassed the Royal Commission then sitting, chaired by Asquith, to consider the affairs of Oxford and Cambridge and their application for state funding for the first time. To the members of this Commission and to their MPs the women and their supporters turned next. However the Commissioners shilly-shallied. The ambivalence of the majority was expressed perfectly in the following recommendation: 'that Cambridge remain mainly and predominantly a "men's University", though of a mixed type'. Only two of the Commission's members, Blanche Athena Clough, Principal of Newnham, and the Labour MP Willie Graham recorded their view that it

> will be very difficult to justify, either in Parliament or in the country, the grant of public money to Cambridge University so long as it refuses to women teachers and students the rights which they now enjoy in every other University in Great Britain.[11]

Nevertheless the legislation which followed the Commission's report failed to reflect the women's lobbying. The new University statutes of 1926 which resulted indeed provided only 'half-a-loaf'. These statutes gave women who took University examinations the titles of decrees – the so-called titular BA, vulgarly abbreviated to the 'BA tit'. They allowed Faculties to recognise the teaching of the Fellows of Girton and Newnham and to give them Faculty membership. They allowed women to present themselves as candidates for University teaching posts. But full membership of the University was still withheld.

II

The women and their colleges remained in this position, trapped on the threshold of Cambridge, until after the Second World War. I want now to explore what such a situation meant socially and academically for undergraduates and senior members: how did it feel to be among but not of, what kinds of social relations, what kinds of work, what kinds of achievements were possible?

At first, at the end of the First World War, social emancipation seemed complete for women undergraduates. Chaperonage could not cope with the flood of returning servicemen and, after its collapse, Frances Partridge recalled, 'we met the men freely, played tennis with them, went punting and on picnics, and above all danced with them. All England had gone dancing mad and so had Cambridge.' The speed of the transition brought its own problems, as she also remembered:

> both sexes were in a far more confusing position than are the permissive generation of undergraduates of today. We were ludicrously inexperienced . . . how did the pleasure got from those attentions connect with love or lust, and how did one know if and when one was what was called with such delusive neatness 'being in love' (as if slipping into a garment)? . . . The confusion only became greater and the colours in the kaleidoscope began to clash when we started to make friends with much more interesting and less amorous males.[12]

The comment is the more telling since Frances had behind her the experience of a coeducational boarding school – Bedales.

Yet as the post-war euphoria subsided the emerging pattern resembled not so much kaleidoscope as world and shadow world. Men and women went their separate ways, the men often largely ignoring the women. An un-named Newnham student expostulated in *Granta* in 1934: 'It is as though the whole University were united in a conspiracy to hush up the existence of women's colleges, which appears to me at once gratuitous and undignified.'[13] The flavour of the society produced by this is very well conveyed in a detective story, *The Cambridge Murders*, by the archaeologist Glyn Daniel, first published in 1945 but set in the Cambridge of 1939. The women characters are either decorative or menial – or both: wives, mistresses, college servants and shop assistants. The work of colleges and of the University and of all the major actors, murderer included, is entirely self-sufficiently male.[14]

A novel with very different ambitions, Rosamond Lehmann's *Dusty Answer*, first published in 1927, gives us a view from inside a women's college. The Cambridge sequences convey a sense almost of a glass wall between men and women students. Within their own institution the lives and relationships of the women students are vivid and intense, in this case with a strong homo-erotic undertone. In public, mixed, university situations the men seem noisy, assured, aggressively three-dimensional, while the women are pale creatures, unsure of their places. Rosamond Lehmann writes of them coming out of the examination halls thus:

> A troop of undergraduates passed on the way from their examina-tion room. They looked amused and exhilarated. They stuffed their papers into their pockets, lit pipes, straightened their shoulders and went cheerfully to lunch.
>
> The girls crept out in twos and threes, earnestly talking, comparing the white slips they carried . . . Girls really should be trained to be less obviously female students. It only needed a little discipline.[15]

Even the collective nouns are distinct: men are undergraduates, women are students.

When the men did take notice of the women students, their attentions were not always kindly in either intent or effect. The astronomer Cecilia Payne-Gaposchkin remembered the experi-ence of being the only woman student attending Rutherford's advanced physics lectures in 1922:

> the regulations required that women should sit by themselves in the front row . . . At every lecture Rutherford would gaze at me

pointedly, as I sat by myself under his very nose, and would begin in his stentorian voice: '*Ladies* and Gentlemen'. All the boys regularly greeted this witticism with thunderous applause, stamping with their feet in the traditional manner, and at every lecture I wished I could sink into the earth. To this day [1979] I instinctively take my place as far back as possible in a lecture room.[16]

Preparing to take her Part II examinations at Newnham in 1923, Cecilia came to the conclusion that there 'was no future for me in England other than teaching'.[17] Her (male) teachers agreed and accordingly she took herself off to the United States. The first doctoral student in astronomy of either sex at Harvard and Radcliffe, she went on to an extraordinarily distinguished career. Recognition came slowly from the Harvard Corporation and the major American observatories tended to relegate women to lowly data-collection activities; there were battles to be fought there too.[18] At the beginning it was clearly second-class citizenship in the republic of letters; nevertheless it was more than Cambridge had offered.

Cecilia Payne-Gaposchkin's decision leads directly into a consideration of the position of the women academics in Cambridge between the wars. They suffered even more than the students from being trapped on the threshold of Cambridge socially, politically and intellectually, in a distinctive and subordinate position, their gender clearly signalled as limiting. The Mistress of Girton and the Principal of Newnham attended University functions and ceremonies by courtesy, not of right. They were, moreover, lumped together with the wives on such occasions, which meant being spectators not actors, wearing conventional hats and gloves, while the academic male peacocks occupied centre stage with scarlet silk robes and black velvet doctors' bonnets. Every academic ritual occasion emphasised the ambivalent position of the academic women: *among but not of.*

This uncomfortable situation was reproduced in matters of University government. The statutes of 1926 had created 183 new

University lecturing posts. To eleven of these women were appointed. Yet they had membership neither of the Regent House nor of the Senate and thus no share in the ultimate determination of the courses that they were appointed to teach. A women who was a Head of Department or a member of a Faculty Board – and by the late 1930s there were a number – could not take part in the public discussion in the Senate House of a report in which her Faculty was concerned. Nor, if the issue subsequently went to a vote of the Regent House, could she cast a vote.

The women responded to this situation in a variety of ways, shaped not only by temperament but by the demands of the discipline they had elected to study. Some did indeed go elsewhere, like Cecilia Payne-Gaposchkin. After the 1921 vote Eileen Power, Girton's brilliant medieval economic historian, wrote that she was 'damned tired of being played fast and loose with by Cambridge University'. She waited neither for the fumblings of the Royal Commission nor for the statutes of 1926 but took herself off to the London School of Economics.[19]

Others, like her equally distinguished historian colleagues at Girton, Gwladys Jones and Helen Cam (who was appointed to succeed her as a medievalist) metaphorically if not literally gritted their teeth and stayed. They served as University lecturers, members of Faculty Boards and devoted teachers of the next generation of students.[20]

In general in the humanities the women could, if they chose, make the 'half-a-loaf' situation palatable. Core teaching was carried out in small groups and remained college-based. Research was a matter of individual endeavour in archives and libraries (although women senior members were not allowed to borrow books from the University Library on the same terms as the men). However when research involved expensive equipment and teams of workers the position of women academics was a great deal more problematic. Between the wars the emergent hybrid discipline of archaeology provided two exceptions that proved this rule. Extraordinarily in the 1930s Newnham boasted not one but two outstanding archaeologists among its research fellows: Gertrude Caton-Thompson, who worked in Egypt and Zimbabwe and Dorothy Garrod, who worked mainly in the Near East and who was in 1937 the first woman to be elected to a Cambridge chair, to the Disney Professorship of Archaeology. The careers of both owed almost everything to private means and impeccable establishment connections outside Cambridge. Dorothy Garrod, whose father was Regius Professor of Medicine at Oxford, took herself to Paris to work with the Abbé Breuil between 1922 and 1924. Gertrude Caton-Thompson worked with Sir Flinders Petrie in Egypt from 1921 to 1926. From 1915 to 1919 she had been secretary to Sir Arthur, later Lord, Salter, ending up with him at the Supreme Economic Council during the Paris Peace Conference. Cambridge needed them, their expertise and connections, much more than they needed Cambridge.[21]

Archaeology was also still taking shape as an academic discipline. In the sciences, in fields where large laboratories, team enterprises and big budgets were already the established order, women academics with or without extraordinary resources and connections had a harder time of it. The career of Constance Tipper demonstrates very clearly what *among, but not of* might mean for a woman scientist. Graduating from Newnham in Natural Sciences in 1915 she did a short period of war work at the National Physical Laboratory, moving then to the Royal School of Mines. Still funded by the Royal School of Mines, she came back to Cambridge in 1923 to work at the Cavendish. With G. I. Taylor she did fundamental work on crystal plasticity in metals. She ceased to be supported by the Royal School of Mines on her marriage to a geologist in 1928; although Newnham came to the rescue with a research fellowship, the Leverhulme Trust subsequently produced some funding and the Engineering Faculty produced a room but not a post.

Only the outbreak of the Second World War really changed things for Constance Tipper. The University Department found it desperately needed her teaching. From 1943 on, she played a pivotal role in the research programme which finally identified the cause of the catastrophic brittle fractures in early welded ships and made life-saving improvements to the design of the 'Liberty' cargo ships on the Atlantic and other cargo routes. The 'Tipper effect' continues to be well known among engineers. Finally in 1949 she was made University Reader in Mechanical Engineering, still the only woman member of the Engineering Faculty.[22]

The Second World War provided the crown of Constance Tipper's scientific career. In this war British women were fully engaged in every theatre in their own right. One of the features which most sharply distinguished Britain from Germany between 1939 and 1945 was the participation of women in the war effort; and not the least of the contributions in Britain was that of women with higher education.

The Cambridge women played their parts to the full. Dorothy Garrod brought her archaeologist's training to bear on the interpretation of aerial photographs for the RAF at Medmenham Abbey. The Newnham mathematician Violent Cane joined the code-breakers at Bletchley.[23] There were Girton and Newnham graduates in almost every government department, including a sizeable contingent at the Board of Trade. In the Shakespearean scholar Muriel Bradbrook and the economist Ruth Cohen these included a future Mistress of Girton and Principal of Newnham respectively. The Oxford chemist Rosemary Murray, who was to become a Fellow of Girton in 1946 and later the founding President of New Hall, the third women's college, began in the Admiralty Signals Department and then climbed the ladder in the WRNS. Teresa Mayor of Newnham and Jean McLachlan of Girton worked for military intelligence, the former specialising in matters French, the latter in matters Spanish. Camilla Wedgwood of Newnham was employed in military intelligence by the Australian Army. Gazetted Lieutenant-Colonel in 1944, she was dropped behind Japanese lines, bringing long experience of anthropological fieldwork in Papua New Guinea to bear in preparing the native populations for the final battles.[24]

III

Eventually and a little grudgingly at the end of 1947 Cambridge fell into line with every other institution of higher education in the country and from 1948 granted women full membership of the University. If this were a folk tale one might now say 'and they all lived happily ever after'; but it would be naive to think that changes in formal rules are always answered by changes in sub-cultures. Fifty years later the transition from 'a men's university though of a mixed type', to a truly mixed university still grinds slowly on. Nevertheless I hope the brief and partial sketch offered here makes it plain that well before women achieved formal membership of the University, while they were *among but not of,* they made a very powerful intellectual contribution, sometimes very much against the odds. And in both wars they contributed notably to the professional and specialist resources of the state.

The range and quality of the women's activities between 1914 and 1918 may also enable us to make some headway with the question of whether the contribution of Cambridge women was distinctive and different in kind from that of Cambridge men – or simply more of the same. In this period surely a great deal of the women's energy went into showing themselves just like the men, into being what the 1990s would call 'male clones', precisely because they were still second-class citizens. The 'half-a-loaf' settlement of the early 1920s represented a very strong pressure on the women to imitate, mimic, duplicate the men. The pressure was far greater than it had been in the pre-1914 years when the women were so far excluded from matters of University policy and government that Newnham at least had felt free to experiment in educational matters.[25] With a foot in the door, how powerful was the moral pressure to conform to the same conventions as the men. How overwhelming must have seemed the case for achieving formal equality before allowing oneself the luxury of exploring the ways in which equal but different might be construed.

Yet it would be a mistake to see the work and achievements of the women in these years simply as a kind of grand shallow play. Gradually, taking infinite pains, they gained first access to and then distinction in a widening range of academic disciplines. In this process they indicated the potential of these disciplines to become gender-neutral in the future. To signal that a discipline might be gender-neutral is not to render it so; but every step that reduces its gendered associations helps. Comparison of the activities of the Cambridge women in the two wars, sketched very partially above, shows how they had quietly enlarged the scope of the contributions they were able to make. They had moved decisively from support to combatant roles. In the Second World War, unlike the First, there was no category of women's war work which could be labelled simply 'substitutes for men'. Nor, apparently, was there any felt need for a separate record of the women's work at the War's end.[26] It stood quite simply on its merits.

NOTES

1 Quoted in Rita McWilliams-Tullberg, *Women at Cambridge* (London, 1975), p. 35.

2 For a fuller discussion of the founders and their aspirations see my essay 'Emily Davies, the Sidgwicks and the education of women in Cambridge' in *Cambridge Minds*, ed. Richard Mason (Cambridge, 1994).

3 McWilliams-Tullberg, *Women at Cambridge*, Appendix A.

4 See Sutherland, 'Emily Davies', pp. 44–5.

5 Vera Brittain, *Testament of Youth* (London, 1933).

6 Leah Leneman, *In the Service of Life. The Story of Elsie Inglis and the Scottish Women's Hospitals* (Edinburgh, 1994).

7 Elsie Butler, *Paper Boats* (London, 1955) p. 68; but see also the remainder of ch. 5.

8 K. T. Butler and H. I. McMorran (eds.), *Girton College Register 1869–1946* (privately printed for Girton College, Cambridge, 1948).

9 Newnham College Archives, 43.

10 The paragraphs which follow are largely based on the admirably full and clear account in McWilliams-Tullberg, *Women at Cambridge*, chs. 9 and 10.

11 *Report of the Royal Commission on the Universities of Oxford and Cambridge*, 1922 Cmd 1588 x, pp. 245, 254–5. For the politics of the Commission, see John Prest, 'The Asquith Commission 1919–1922' in *The History of the University of Oxford*, vol. VIII: *The Twentieth Century*, ed. Brian Harrison (Oxford, 1994).

12 Frances Partridge, *Memories* (London, 1981), pp. 62–3.

13 *Granta*, 16 May 1934, 'What Cambridge women want' by a Newnham student.

14 It was still being reprinted by Penguin Books in the 1960s.

15 Paperback edition, Harmondsworth, 1936, p. 185. Rosamond Lehmann read English and Modern Languages at Girton from 1919 to 1922.

16 Katherine Haramundanis (ed.), *Cecilia Payne-Gaposchkin: An Autobiography and Other Recollections*, 2nd edition (Cambridge, 1996), p. 118.

17 Ibid. p. 124.

18 Ibid. pp. 25–7.

19 Maxine Berg, 'Eileen Power 1889–1940' in Carmen Blacker and Edward Shils (eds.), *Cambridge Women: Twelve Portraits* (Cambridge, 1996), direct quotation from p. 171. See also Maxine Berg, *A Woman in History: Eileen Power, 1889–1940*, (Cambridge, 1996).

20 For Helen Cam, see Janet Sondheimer's essay in Blacker and Shils (eds.), *Cambridge Women*. M. G. Jones awaits a biographer; but some sense of her quality can be gathered from a reading of her classic study, *The Charity School Movement: A Study of Eighteenth-Century Puritanism in Action* (Cambridge, 1938).

21 Gertrude Caton-Thompson, 'Dorothy Garrod 1892–1968' *Proceedings of the British Academy*, 55 (1969), pp. 339–61 and *Mixed Memoirs* (Gateshead 1983). See also Colin Renfrew, 'Three Cambridge prehistorians' in Mason (ed.), *Cambridge Minds*.

22 Constance Tipper, *The Brittle Fracture Story* (London, 1962); Jim Charles and Gerry Smith, 'Constance Tipper: her life and work' *Materials World* 4, 6 (June 1996), pp. 336–7.

23 Personal information. For Bletchley, see *Codebreakers. The Inside Story of Bletchley Park*, ed. F. H. Hinsley and Alan Stripp (Oxford, 1993).

24 *Newnham College Roll Letter 1956* (Cambridge, privately printed, 1956). When no other source is cited, biographical information is drawn from the Girton College and Newnham College *Registers*. The war service of those named is a tiny fraction of the whole.

25 Sutherland, 'Emily Davies', pp. 37–40.

26 Above, pp. 90-1.

CHAPTER 6

Cambridge Classics for the third millennium

PAUL CARTLEDGE

The dread words 'Classics' and 'Classicist' tend, I suspect, to conjure up an image not just of the antique but of the antiquated: of dryasdust pedants poring over time-expired creations of literary or visual art. That image is of course a million miles from the reality. In this essay, therefore, rather than only looking back, I shall be also – and chiefly – looking forward, taking a leaf out of *Foundations for the Future*, the catalogue accompanying Cambridge University's pioneering exhibition of that title held at Christie's early in 1995.[1] I shall not neglect the foundations of the present-day study and teaching of Classics in Cambridge, but the emphasis will be placed firmly on their future in the foreseeable term.

My main theme will be cross- or inter-disciplinarity. Classics will be represented here as above all a systematically interdisciplinary practice. The Faculty of Classics is divided into six Groups (also known as Caucuses).[2] Five are more or less traditional: Language and Literature, Philosophy, History, Archaeology and Art History, Philology and Linguistics. But the sixth and latest, instituted only in the 1980s, is Interdisciplinary ('Group X'). Courses offered under this latter rubric not only borrow something from all or most of the other five, but also add something new of their own, namely cross-disciplinary links between Classics and adjacent subjects, especially the literary and social science disciplines such as Archaeology and Anthropology, Architecture and Art History, English, History, Modern and Medieval Languages, Philosophy, and Social and Political Studies. Examples of Classics's 'foundations' will be drawn below from each of the five more traditional subdisciplines, but the second and more important part of this essay, on the future of Classics, will take its illustrations almost entirely from the Interdisciplinary Group.

FOUNDATIONS

I should like to begin with a personal observation. As I cycle in of a morning to the centre of Cambridge from Trumpington on the south where I live, I pass on my right-hand side the following streets: Porson Road, Barrow Road, Bentley Road, and Newton Road. All are named after Trinity men, but it is not their college affiliation that interests me now, but the fact that that they were all in one sense or another Classicists.[3] Isaac Barrow (1630–77) is best known as a mathematician, and more especially as the teacher of another mathematician named Isaac, a certain Newton. But before Barrow became the University's first Lucasian Professor of Mathematics in 1663 (inaugurating the chair held today by Stephen Hawking), he had been appointed in 1660 to the Regius Chair of Greek founded by Henry VIII (a chair now held by P. E. Easterling, its first woman holder). Elegantly combining his Maths with his Classics, Barrow produced Latin versions of both Euclid and Archimedes, but in 1669, at the grand old age of thirty-nine, he resigned from the Lucasian chair to make way for Newton, a dozen years his junior (1642–1727).

By 1684 Newton had demonstrated the whole gravitation theory, and expounded it – in Latin – in *De Motu corporum* ('On the Motion of Bodies'). Three years later – again in Latin – he published his most celebrated single work, the *Principia*, or to give it its full title *Philosophiae Naturalis Principia Mathematica* ('The Mathematical Foundations of Natural Philosophy'). It came blessed with a prefatory dedication to Newton, in competent Latin hexameters, by Edmund Halley, and was published under the auspices of the Royal Society, the then President of which was Samuel Pepys. Small world. But I must insist on the fact that it was a world of which the scholarly *lingua franca* was Latin. Newton's major works were thus written, and thought, not only in English but also importantly in Latin. Newton, however, as is well known, was not the very model of a modern scientist as we understand that term today; he more than just dabbled, for instance, in alchemy. What is not so well known, perhaps, is that to his dying day he also more than dabbled in ancient history, and more specifically in ancient chronography, taking such an enlarged view of 'ancient' as might shame many of his more parochial epigones in the field. His last published work, which appeared posthumously in 1728, was entitled *The Chronology of Ancient Kingdoms Amended.*[4]

Yet Barrow and Newton are not considered 'Classicists' today, fairly enough. My other two street eponyms, however, certainly are; indeed, they are universally accounted among the founding fathers of my discipline. Richard Bentley, a Yorkshireman, was a younger contemporary of Newton (1662–1742). In his day, he was as famous or notorious for the succession of scandals, scholarly or otherwise, in which he chose to get involved. Suffice it to say that as Master of Trinity in the 1720s he even had to survive a trial and attempted ejection by the Vice-Chancellor. But in the longer perspective it is Bentley's philological scholarship that has justly earned him ancestor worship. He published critical texts of many classical authors, including the authors of the Greek New Testament. Not the least of his achievements in the publishing sphere was to help reorganise the University's Press, which remains noted to this day for a Classics list that includes – any selection would be invidious – the series of 'Cambridge Greek and Latin Classics' editions of texts, and the *Cambridge Ancient History.*

Fourthly, and finally, Richard Porson (1759–1808) was appointed Regius Professor at the tender age of thirty-three. Like

Bentley, Porson distinguished himself as a textual critic, making his name in the 1780s in the controversy over the authenticity of a passage in the First Epistle to John (chapter 5, verse 7). He went on to edit four plays of Euripides, thereby unconsciously anticipating another Cambridge Classical tradition that is still going very strong indeed, the Cambridge Greek Play (staged every three years, in Greek, since the 1880s, with music sometimes written by such distinguished figures as Charles Villiers Stanford, Hubert Parry and Ralph Vaughan Williams – the latter for Aristophanes's *Wasps*). Yet Porson, sad to record, was not only (or best) known to his contemporaries for his scholarship. He was also a legend in his own lunchtime, as we might rather coarsely say, for his crapulence. I quote from a letter by a former Trinity undergraduate:[5]

> I remember to have seen Porson at Cambridge, in the hall of our college, and in private parties, but not frequently; and I never can recollect him except as drunk or brutal . . . He was tolerated in this state amongst the young men for his talents, as the Turks think a Madman inspired, and bear with him. He used to recite, or rather vomit pages of all languages, and could hiccup Greek like a helot.

That missive was sent from Venice in 1818, ten years after Porson's death, to the publisher John Murray. It was written by Lord Byron, who thereby incidentally showed himself to be something of a thinker (as well as a drinker) himself. The helots he mentions were the native Greek servile underclass whom the mighty Spartans callously exploited in several ways. One of these involved making helots paralytically drunk on unmixed wine and then exhibiting them in that enforced condition as a sort of freak show in the officers' mess, in order to teach their sons a lesson in what was not the proper way for free Greek men to behave in 'polite' company.

Porson and Byron take us, just, into the nineteenth century. This is the century in which Cambridge Classics as we know it today – a Tripos or university degree course – was first properly taught and examined, from 1824. It was only the University's second such Tripos, although it came a long way after Mathematics in both chronology and status.[6] Around the middle of the century, in 1859, one Alfred Edward Housman was born, not in Shropshire of course but in Worcestershire. Housman became perhaps the foremost Latinist of his day, at any rate in the English-speaking world, and that despite failing to pass the degree course which I myself took a century later, 'Greats' (Ancient History and Philosophy) at Oxford. In 1911 Housman's pre-eminent Latinity was at last duly recognised when he was appointed to our Kennedy chair. To the world at large he is known as a poet. But to the classical *cognoscenti* he is as famous or notorious for his critical edition of one of the more obscure ancient Latin authors, Manilius, who in the first century AD composed a long poem in hexameters on astrology. In 1931, when in his anecdotage (he died in 1936), Housman had the ill luck to encounter Harold Nicolson, husband of Vita Sackville West and diarist *extraordinaire*, who penned the following less than flattering portrait:[7]

> Housman is dry, soft, shy, prickly, smooth, conventional, silent, feminine, fussy, pernickety, sensitive, tidy, greedy, and [the sting in the tale] a touch of a toper.

Drawing a veil swiftly over the apparent penchant of Classicists for the demon rum (or more accurately the divine juice of Dionysus), I pass in celebration to the posthumous three-volume edition of the Housman papers, one of whose editors was my senior colleague, Professor James Diggle.[8] It is in another of his capacities, however, that I want more especially to commemorate this phenomenally learned scholar of our own day. Diggle was from 1982 to 1993 the University's Orator, one of whose functions it has been since the office's establishment in (probably) 1521 to write a congratulatory address – still in Latin, but now with an attached English translation or version – on behalf of those elite few men and women whom the University deems worthy of an honorary degree. Strangely enough, very few of the honorands have been Classicists. Not so strangely, many of them have been scientists, working in fields for which ancient Latin does not often automatically possess a suitable vocabulary.

Just how adroitly Professor Diggle regularly triumphed over that linguistic difficulty may be seen from the selection of fifty of his orations published in 1994.[9] The honorands included in this minor masterpiece range from David Attenborough to Elisabeth Frink, from Mary Leakey to His Majesty Juan Carlos I King of Spain, from Jacques Derrida to Nadine Gordimer. It is, however, from Diggle's honorific address for my fellow-lecturer Maurice Wilkes, Emeritus Professor of Computer Technology, that I wish to quote the following snatch, partly because it takes us back to Barrow and Newton, but mainly because it brings Classics into contact with the very latest revolution of practical intellectual wisdom:

> Annis abhinc centum et quinquaginta Carolus Babbage, Mathematices Professor Lucasianus, organum ἀναλυτικὸν excogitauit, computatricis machinae quae nunc est proplasma, sed pro materiae copiis quae tunc erant praematurius. adest qui Caroli inuento summam manum imposuit, quippe qui ultimum illud quod in uotis manebat, memoriae facultatem, inuenerit. nempe praecepta machinae tradenda taeniis impressit impressaque in sonos ultra aurium captum acutos conuertit. hos in fistulam hydrargyro refertam inmisit et hydrarguri densitate tardatos in unum perpetuum diaulum circinauit. en nouum productum miraculum, machinam memori mente praeditam, laudibus paene Vergilianis efferendam:
>
> > spiritus EDSAC alit totamque infusa per artus
> > mens agitat molem et magno se corpore miscet.

Which being translated (approximately, and wittily, by Diggle himself) means:

> A century and a half ago Charles Babbage, Lucasian Professor of Mathematics, devised his Analytical Engine, which embodied most of the concepts which we now take for granted in the digital computer ['computatrix machina']. But those concepts were far

ahead of the available technology. Here stands the man who finally brought to its fullest reality Babbage's dream, by providing the computer with the one vital organ which it still lacked: a capacious memory. He converted a program and its data, punched onto paper tape, into ultrasonic pulses, and he fed the pulses into a tube of mercury, which delayed their progress, and he caused the pulses to circulate indefinitely. And so EDSAC was born, the Electronic Delay Storage Automatic Calculator, the first fully operational computer with its own memory store. As Virgil almost said [in his Latin version, Diggle replaced Virgil's 'intus' = 'within' by 'EDSAC', scanned in accordance with the normal rules as a trochee],

> A spirit nourishes the parts within,
> And mind moves matter in the mighty frame (*Aeneid* 6.726–7)

It would be difficult, you might perhaps think, to find anyone in this century whose facility in Greek was the equal of Diggle's or Housman's in Latin. But in Cambridge we are spoilt for choice. If choose I must, I would cast my vote for John Chadwick. To not many scholars is it given both to decipher a script and in so doing to increase at a stroke the known chronological range of a language by something approaching a thousand years. That nevertheless is what Chadwick achieved, working closely with the amateur cryptographer (and professional architect) Michael Ventris. Together they proved that the prosaically named Linear B script, applied to clay tablets excavated in a handful of palace sites in Crete and mainland Greece, had been used to transcribe an early form of the Greek language. They thereby pushed the known timespan of Greek back from about 700 BC to about 1500 or earlier. Chadwick could not have done what he did, the crucial basic philology, without Ventris's intuitive brilliance, but neither would Ventris have been able to develop his intuitions and persuade the scholarly community of their validity without Chadwick's linguistic scholarship.[10]

I have almost finished with 'Foundations', which are essentially linguistic and philological. I am almost ready to move on to the 'Future', which in my presentation will be essentially cultural-historical. But as a bridging transition between the two I introduce a scholarly trinity whose work, often consciously interrelated and sometimes actually collaborative, looks forward to the present and future of Cambridge Classics. Though philologically based, as all Classical scholarship must ultimately be, it nevertheless moved decisively beyond the sphere of philology in any narrow sense to make vital inroads into adjacent disciplines, especially the then newborn Anthropology. My trinity consists of J. G. Frazer (1854–1941), F. M. Cornford (1874–1943), and – ladies last, as too often in Cambridge – Jane Ellen Harrison (1850–1928).

Frazer, according to the late Ernest Gellner, was 'certainly the most famous British anthropologist, and quite probably the most famous anthropologist altogether'.[11] But he was also a Classicist, and just a few years after publishing the first edition of *The Golden Bough* (the title of which he had borrowed from Virgil) he published in 1898 an excellent six-volume edition of the ancient travel writer Pausanias. Pausanias, a Greek of the second century AD from Asia Minor, was a cultural tourist and a sort of retrospective ethnographer of Greek religion, long before the terms ethnographer or anthropologist were coined or could have been imagined; in other words, he was a sort of proto-Frazer. Frazer subsequently edited also the 'Library' of the late Classical mythographer Apollodorus. His major work, nevertheless, was done outside the now ever more strictly delimited Classical field.

Not so that of Francis Cornford, whose career culminated in his election as Laurence Professor of Ancient Philosophy in 1931. All the same, Cornford too is probably more widely known today for a non- or meta-academic work, his little squib of a pamphlet entitled *Microcosmographia Academica* (first published in 1908).[12] This is marketed now (though not by the Cambridge University Press) as 'Cambridge's Classic Guide to Success in the World', but Cornford himself had far more limited targets and aims in view, directing his satire at what he saw as the diehard conservatism of

Cambridge University politics. Part of the reason for its enduring success is that Cornford was particularly adroit at giving ironic labels and definitions to the conservative, or reactionary, ploys used then (and still) by academic politicians of the right: there is, for example, the 'Principle of the Wedge', according to which 'you should not act justly now for fear of raising expectations which you are afraid you will not have the courage to satisfy'.

It is not, however, as an academic satirist that Cornford principally interests me here, but as a would-be reformer of Classics, when – not for the last time – 'the subject was under threat, not just in the Universities, but in the schools and in the minds of all literate people';[13] and as a key figure in what is now generally referred to as the 'Cambridge Ritualist School' of scholars of ancient Greece.[14] The latter were Classicists inspired by the great growth of ethnography as a discipline in the late nineteenth century, an epoch when Herodotus could still serve as the ultimate founding father for both Classics and anthropology before the latter broke sharply with the former. These pioneers therefore sought to study the ancient Greeks, not as a sort of superhuman species of divinely rational beings on the lines of the Victorians' dominant stereotype, but more as if they were a sort of contemporary 'primitive' tribe equipped with all kinds of funny, irrational customs, especially in the religious sphere of myth and ritual.

An even more key member of that School was Cornford's teacher and mentor, Jane Ellen Harrison of Newnham, who deservedly wins a place in a recent publication entitled *Cambridge Women: Twelve Portraits.*[15] Harrison was first a student and later a Fellow of Newnham, in the period when women were first being permitted an active place, but with very much a second-class status, in male-dominated Cambridge academic life. In fact, Cambridge has the unfortunate distinction of being the last university in England to grant women equal status with men, and they could not take a full degree until 1948, a mere twenty years after women in the United Kingdom had achieved full adult suffrage.[16] Perhaps it was understandable that, as a woman, Harrison should have been attracted more towards the side-alleys and the dark subterranean channels of ancient Greek society, and especially to the areas of religion in which women could play a significant role, and away from the sundrenched, politically enfranchised public spaces occupied exclusively by Greek men, who, however, wrote almost all the literature and other texts and indeed created almost all the extant artefacts that have enabled Harrison (and us) to study the Greeks.

Not everyone approved of the Cambridge Ritualist School then, any more than Cambridge Classics wins universal approval today. One rather conservative and puritanical American scholar, indeed, went so far as to denounce 'the anthropological Hellenism of Sir James Frazer, the irrational, semi-sentimental, Polynesian, free-verse and sex-freedom Hellenism of all the gushful geysers of "rapturous rubbish" about the Greek spirit'.[17] That was many decades ago, but that hostile critic's spiritual descendants are with us still. Nevertheless, in so far as we here in Cambridge try to keep the Classical flag flying by constantly renewing our discipline from within and without, these three are the sort of ancestors I think we should select to honour for the future.

THE FUTURE

Cultural heritage, however, unlike natural heredity, is not a matter of Darwinian evolutionary process. For the Cambridge School of Cornford and Harrison to spawn significant offspring several further conditions had to be satisfied, not least among them being the appearance of a sort of *deus ex machina*. That was duly arranged for us in the mid-1950s, far away from Cambridge on the other side of the Atlantic, by US Senator Joseph McCarthy. The divine (or at any rate heroic) apparition took the shape of Moses, later Sir Moses, Finley (1912–86), who arrived opportunely at Jesus College in 1954.

It is impossible to sum up Finley's achievement in a few words.

Besides helping to transform ancient history among specialists – turning it away from events and narrative and preoccupation with politics to structures, ideas and sociological analysis – he was also the most powerful apostle of his day for Classics in British society at large, both in the schools and the multiplying universities and, through his radio broadcasting and magazine articles, among the wider general public. Classics in Britain in the 1950s and 1960s was beginning to suffer badly from the white heat of the technological revolution. It could be portrayed (as it sometimes still is), quite unfairly, as simply irrelevant to the needs of the modern, science- and technology-based world. Worse, it carried with it the twin stigmata of elitism and failure: elitism, in that the study of Latin and Greek to a high level at the secondary and tertiary levels of education had always been the privilege and preserve of the political, social and economic as well as intellectual elite; failure, in that the deprivations and devastations of two world wars could be somehow blamed collectively on a ruling and governing class formed in the main by that perniciously elitist and irrelevant Classical education.

Finley thus found himself, like Cornford, having to contend with a rising tide of anti-Classicism, a 'crisis in the Classics'.[18] The dropping of Latin as a requirement for entry to Oxford in the 1960s was just the most obvious sign of the hostile times. Far more serious, though, was the widespread jettisoning of Latin, not to mention ancient Greek, as secondary school subjects. Finley's response, ably abetted by colleagues such as Cambridge's Kennedy Professor Charles Brink, was to promote the founding of the Joint Association of Classical Teachers (JACT). This was envisaged both as a defence taskforce, uniting teachers at the secondary and the tertiary levels, and simultaneously as a weapon of attack, devising new curricula and new ways of teaching the Classical languages from scratch at university as well as or rather than at school.

Hence, first, the 'Cambridge Greek Course', then more recently the 'Cambridge Latin Course', both now used worldwide – and further developed outside Cambridge. Hence, too, a new Ancient History course, JACT-developed and JACT-run, for school pupils in their years immediately before university. This was and is a course that stresses economic, social and cultural history more than old-style drum-and-trumpet, kings-and-battles political-event history; and that does so unashamedly through English translations of the ancient texts as well as or in the stead of the supposedly ancient-text only study of the historical sources in old-style Classics.[19]

Finley's own specialist publications, after his pioneering applications of sociology to Classical Athenian landownership in 1952 and of anthropology to the world of Homer in 1954, took the form of articles rather than books, and throughout the later 1950s and 1960s a steady stream of articles on ancient Graeco-Roman society and civilisation poured from his pen. Unflinchingly, Finley placed great emphasis on the seamiest side of ancient civilisation, on the fact that it was somehow dependent on chattel slavery – that is, on treating hundreds of thousands of human beings as if they were less than human, as mere items of property. Here, I think, it made a very great difference that Finley was American, since the best scholarship on slave systems anywhere in the world was at that time being practised in North America, in particular on the 'peculiar institution' of chattel slavery in the American Old South.

But just being American by birth and training would not have enabled Finley to make the best possible use of that research material. For that, he needed also a theory of history and the sophisticated awareness of historical methodology that he had acquired by working during the 1940s at the Institute for Social Research in New York. This was the forcibly transplanted version of the so-called Frankfurt School of social theorists, expelled from its European home by Nazi racism. In particular, Finley applied the comparative method to ancient Greek and Roman history, and not only to the history of ancient slavery but also to other economic and non-economic aspects of antiquity, using comparison skilfully to highlight differences as well as similarities in structure, organisation, and ideology between one society's arrangements and another's. Since Finley also wrote so well, it is a pleasure to report that all his major work is still currently in print, and almost all of it in paperback.[20]

It is this powerful legacy – of comparativism in method and preoccupation with society, economy and ideology in Greece and Rome – that we Cambridge Classicists do and should continue to seek to develop and enhance, as we face yet more if rather differently focused attacks on our discipline. I mention just two such currently ongoing assaults. First, there are the so-called Canon or Culture Wars, a mainly but not by any means exclusively North American phenomenon. According to the anti-Canonical culture-warriors, almost all the texts studied in Classics, the mother of all Canons, exhibit the excessive influence of 'Dead White European Males', who, they aver, should be at the very least dethroned, if not decapitated, to make way for the study of living, non-white, non-European, non-male authors and creators.[21]

The other major assault on Classics, also chiefly North American, although its best known exponent is ironically not only British but a Cambridge man (from King's), targets the modern tradition of Classical scholarship rather than the status of its canonical authors. The charge in this case is that Classical scholarship since the middle of the eighteenth century has been fatally tainted by a sort of retrospective racism or ethnic cleansing, the aim or at any rate the effect of which has been to deny to the non-Greek and non-Roman peoples around the Mediterranean their rightful place in the creation of the western cultural tradition. Particularly badly treated, it is alleged, have been the Egyptians and the Phoenicians, from whom the Greeks and Romans borrowed crucial components of what they – or at any rate many modern Classicists – have passed off as their own original inventions.[22]

The proper way for us to take these would-be knockout

punches is, I think, on the chin, by engaging with the critics on
their own terms and ground. In the first instance, we should join
in reconsidering the canon from the point of view of why and
how certain works of visual and literary art became canonical
(and why others did not). We here in Cambridge are in a
particularly fortunate position to do so. We have our Fitzwilliam
Museum, stuffed with genuine and often lovely antiquities, some
of them true 'treasures' .[23] And we have our Museum of
Classical Archaeology (the 'Ark'), housed together with our
Faculty Library in a new, purpose-designed Faculty of Classics
building on the Sidgwick Site. The 'Ark' contains not only a
small selection of antiquities but also the second-best collection
of plaster casts of antiquities in the world, an invaluable teaching
resource. The circumstances in which the casts were acquired
and exhibited themselves form an integral part of the sort of
reflexive study of Classics and the Classical that I am here
advocating.[24]

In the same vein, the best possible way to respond to the charge
of Classical cultural chauvinism, ethnocentrism, or even racism is
to follow the lead of my colleagues Mary Beard and John Henderson in their *Classics: A Very Short Introduction*.[25] That is, instead
of putting the Greeks and Romans up on a pedestal for us to gawp
at in boundless admiration, they ask why it is that we — that is,
those of us with a broadly European cultural formation — have
since the Renaissance chosen to find them admirable and, up to a
point, imitable. They show, moreover, in a brief compass and in
an anthropologising sort of way, just how yawningly broad the
gulf may be between such idealising reception and the sometimes
altogether different realities of ancient Graeco-Roman life,
culture and civilisation. Nor is this approach of theirs, that is ours,
confined to published books. It is related to, indeed based upon,
our regular teaching of undergraduates, as exemplified in the
recent Group X course entitled 'Classics: Nineteenth- and Twentieth-Century Perspectives'.

Such reflexivity does admittedly have its costs. The ancient

Greeks and Romans tend to come out of the exercise with their image looking significantly less glamorous, even perhaps a shade tarnished, compared with the way it looked, say, fifty years ago, when E.V. Rieu founded the Penguin Classics library and inaugurated it with his best-selling translation of the *Odyssey*.[26] A nice measure, in fact, of the profound changes in cultural sensibility and sensitivity that have occurred between 1946 and the 1990s, both in our western or western-influenced societies and consequently in attitudes to the Graeco-Roman legacy, is provided by the sharp cultural, not just formal, gap between Rieu's prose *Odyssey* and the poem-cycle *Omeros* (1990) by Nobel prizewinning poet-playwright Derek Walcott. *Omeros* (so titled after the modern Greek pronunciation of 'Homer') is a transposition and transformation – a translation in quite another sense – of Homer's epic, and as I suggested in an article on the place of slavery in ancient Greek society and in modern western imaginations, the Classical tradition, with Homer's *Odyssey* to the fore, seems to have served the slave-descended West Indian Walcott as 'an irritant, something like the grit in an oyster that eventually produces a pearl'.[27]

It is thus with a quotation from *Omeros* that this essay may most appropriately end, to illustrate both the continuing vitality of the Classical tradition and – therefore – the continuing need for us Classicists to be in a position to interpret the ever-changing tradition in the light of its original ancient sources in the world of Greece and Rome:

> 'Somewhere over there', said my guide, 'the Trail of Tears
> started.' I leant towards the crystalline creek. Pines
> shaded it. Then I made myself hear the water's
>
> language around the rocks in its clear-running lines
> and its small shelving falls with their eddies, 'Choctaws',
> 'Creeks', 'Choctaws', and I thought of the Greek revival
>
> carried past the names of towns with columned porches,
> and how Greek it was, the necessary evil
> of slavery, in the catalogue of Georgia's

marble past, the Jeffersonian ideal in
plantations with its Hectors and Achilleses,
its foam in the dogwood spray, past towns named Helen,

Athens, Sparta, Troy.

NOTES

I am indebted to Sarah Ormrod for inviting me to address the members of
the International Summer School, and to the (to me) surprisingly large
number who attended my talk, especially those who came up afterwards
with both questions and personal reflections. Pat Easterling with her usual
acuteness saved me from needless error and with her usual generosity
suggested further amendments. The title of the original talk was 'Cambridge Class*ix* for the Third Millennium', the '-ix' suffix being chosen
partly for its impeccable Latinity but mainly because 'X' in the jargon of
the Classics Faculty refers to Group or Caucus X (see above, p. 103).

Bibliographical note: except where otherwise specified, all the works
cited below were published by Cambridge University Press, the world's
oldest publisher in continuous operation to the present day: see M. H.
Black, *Cambridge University Press, 1584–1984* (1984).

1 Sir James Holt (ed.), *Foundations for the Future: The University of Cambridge* (1995). Most aspects of the University's past and present are at
least touched upon, often humorously, in Elisabeth Leedham-Green,
A Concise History of the University Cambridge (1996).

2 See *Prospectus: Cambridge Classical Courses* (latest edition 1997/98, now
produced 'in-house'). Within my own time at Cambridge (since 1979)
the Faculty has lost its former Group F (for Roman Law), but happily
the University maintains its Regius Professorship of Civil Law, the
present incumbent of which, David Johnston, is a Classicist by
training and retains strong ties with Classics.

3 Anecdotal and other information about these and many other Cambridge luminaries is gathered and accessibly presented in Laurence
and Helen Fowler (eds.), *Cambridge Commemorated: An Anthology of
University Life* (1984 and reprints).

4 R. Westfall, *The Life of Isaac Newton* (1993; Canto edn, 1994) esp. pp. 300–2.

5 The Fowlers, *Cambridge Commemorated*, p. 132.

6 Frank Stubbings, *Bedders, Bulldogs and Bedells: A Cambridge Glossary* (2nd edn, 1995) s.v. 'Tripos'. Dr Stubbings is himself a Classicist and prehistorian. A proper scholarly history of the Classical Tripos is currently being undertaken by Dr Chris Stray; meanwhile see his 'Contestation and change in Cambridge Classics, 1822–1914', *DIALOGOS* 4 (1997), 95–109.

7 The Fowlers, *Cambridge Commemorated*, p. 303.

8 J. Diggle and F. Goodyear (eds.), *The Classical Papers of A. E. Housman*, 3 vols. (1972). See also C. O. Brink (a late holder of the Kennedy Chair of Latin), *English Classical Scholarship: Historical Reflections on Bentley, Porson and Housman* (James Clarke, 1986). Sir Tom Stoppard's scholarly dramatised re-creation of Housman, *The Invention of Love*, opened at the Royal National Theatre in October 1997.

9 J. Diggle, *Cambridge Orations 1982–1993* (1994); his lively introduction (pp. ix–xvi) discusses the origins and functions of the office. The Latin oration and its English version from which I quote below are printed on pp. 100–1.

10 J. Chadwick, *The Decipherment of Linear B* (2nd edn, 1967; Canto series reprint with 'Postscript', 1992). Chadwick's work has been further developed in Cambridge by my senior Classics colleague John Killen, a graduate of another distinguished and ancient Trinity – Trinity College Dublin (founded 1592).

11 E. Gellner (formerly Cambridge's Wyse Professor of Social Anthropology), 'James Frazer and Cambridge anthropology', in Richard Mason (ed.), *Cambridge Minds* (1994), pp. 204–17, at 204.

12 Cornford's *Microcosmographia Academica* (1908) is best read in G. Johnson, *University Politics. F. M. Cornford's Cambridge and his Advice to the Young Academic Politician* (1994).

13 Johnson, *University Politics*, p. 61. See now C. A. Stray, *Classics Transformed: Schools, Universities and Society in England, 1830–1960* (Oxford University Press, 1998).

14 W. M. Calder III (ed.), *The Cambridge Ritualists Re-Considered* (Scholars Press, 1991).

15 E. Shils and C. Blacker (eds.), *Cambridge Women: Twelve Portraits*

(1996); this includes also Eleanor ('Nora', née Balfour) Sidgwick, Newnham's effective joint founder – see next note; and Frances (née Darwin) Cornford, Francis's wife and Charles Darwin's grand-daughter.

16 G. Sutherland, 'Emily Davies, the Sidgwicks and the education of women in Cambridge', in *Cambridge Minds*, pp. 34–47.

17 Paul Shorey, as quoted in C. Kluckhohn, *Anthropology and the Classics* (Brown University Press, 1961), p. 20.

18 Finley, 'Crisis in the Classics', in J. H. Plumb (ed.), *Crisis in the Humanities* (Penguin, 1964), pp. 11–23.

19 These courses and their associated coursebooks stem from the Cambridge School Classics Project, on which see M. Forrest, *Modernising the Classics. A study in curriculum development* (University of Exeter Press, 1996). JACT-related publications include *The World of Athens* (1984) and P. E. Easterling and J. V. Muir (eds.), *Greek Religion and Society* (1985). One reflection of the current growth of interest in studying the ancient world is that the Open University has now (autumn 1996) begun to teach beginners' Greek; the initial enrolment numbered some 500 students.

20 M. I. Finley's works in print (Penguin, except where otherwise noted) include *The World of Odysseus* (Viking Press 1954; 2nd edn, 1977); *The Ancient Greeks* (1963); *Aspects of Antiquity* (1972, 1977); *The Use and Abuse of History* (Chatto & Windus, 1975; Hogarth Press, 1986) (this includes his 1971 Cambridge Inaugural Lecture and his 1972 Jane Harrison Memorial Lecture at Newnham College); *Economy and Society in Ancient Greece* (Chatto & Windus, 1980; 1983); *Politics in the Ancient World* (Cambridge University Press, 1983); *Ancient History: Evidence and Models* (Chatto & Windus 1985; 1994).

21 B. M. W. Knox's *The Oldest Dead White European Males and other reflections on the Classics* (Norton, 1993) is not an entirely satisfactory rejoinder: see my review in *The Times Higher Education Supplement*, 3 December 1993, p. 21.

22 Martin Bernal's deliberately provocative *Black Athena. The Afro-Asiatic Roots of Western Civilization* (2 vols. to date, Free Association Press, 1987–91) almost inevitably provoked a conservative counterblast, Mary Lefkowitz's *Not Out of Africa. How Afrocentrism Became an Excuse to Teach Myth as History* (Free Press, 1996). Neither of these extreme

stances is entirely helpful, the former because it has lent itself unfortunately to racialist interpretations, the latter because it seeks to defend 'truths' that are themselves open to and in constant need of critical reassessment.

23 *Treasures of the Fitzwilliam Museum* (Pevensey Press, 1982). Richard, seventh Viscount Fitzwilliam (1745–1816) bequeathed to the University his library and fine art collection; for just one facet of his taste and philanthropy, see the article by C. Ballinger (herself a Cambridge Classics graduate), *Early Music Today* (Aug./Sept. 1996), 9–13.

24 M. Beard, 'Casts and cast-offs: the origins of the Museum of Classical Archaeology', *Proceedings of the Cambridge Philological Society* 39 (1993), 1–29.

25 M. Beard and J. Henderson, *Classics: A Very Short Introduction* (Oxford University Press, 1995).

26 P. Cartledge, 'So different and so long ago: 50 years of Penguin Classics', *New Statesman & Society*, 1 March 1996, pp. 36–7.

27 P. Cartledge, '"Like a worm i' the bud"? A heterology of Classical Greek slavery', *Greece & Rome* 40 (1993), 163–80. See also Cartledge, *The Greeks. A Portrait of Self and Others* (Oxford University Press, new edn. 1997) esp. ch. 6; and P. Cartledge (ed.), *The Cambridge Illustrated History of Ancient Greece* (1997).

CHAPTER 7

Cambridge contributions: the philosophy of science

PETER LIPTON

To admit at a cocktail party that one does philosophy of science is a good way to end the conversation. Many people have only the haziest idea what philosophers do and many people think that philosophy and science have nothing to do with each other. So I will begin with some general remarks about the philosophy of science, before turning to the great Cambridge tradition in the subject. Finally, because the only way properly to appreciate philosophy is to worry a philosophical problem for oneself, I will present a puzzle about the way scientists test their theories.

JUSTIFICATION AND DESCRIPTION

What is the philosophy of science? The subject can be seen to emerge from more general areas of philosophy. One of the most important of these is epistemology, the theory of knowledge. The central issues are what knowledge is, how much of it we have, and how we acquire it. Epistemology often proceeds by presenting very negative, destructive arguments, arguments that seem to show that we do not know what we think we know, arguments that seem to show that we know almost nothing. Some of these sceptical arguments are familiar to adults and often reinvented by children. For example, you might worry that the distinctive experiences you have while, say, riding a bicycle, could in fact be experiences that you are having in the comfort of your bed. It could all be a dream that feels just like riding a bike, but none of it would be real. Given that these two situations – the real bicycling situation and the dream situation – feel, seem, exactly the same, on what rational basis do you believe the bike hypothesis over the dream hypothesis? That is the sort of argument that children and epistemologists worry about.

To understand better what knowledge is, how it works and how it changes, it helps to think about specific types of knowledge. If you made a list of the different types of knowledge we claim to have, scientific knowledge would probably come near the top. Science seems an example of knowledge acquisition at its most articulate, most ambitious and, many would say, most successful. So if one wants to understand knowledge, it pays to have a good look at science. That makes the connection to the philosophy of science, since much of the philosophy of science is the theory of knowledge with science as the example. We are trying to understand how scientific knowledge develops, how it changes. How does science work, how do scientific theories get produced, how are they tested, how are they evaluated, how do scientists weigh evidence? Those are the sorts of questions that one asks in the philosophy of science. They are not the only questions philosophers of science ask – there are lots of others – but they form a central part of the discipline.

Philosophical questions about scientific knowledge fall into two groups. In the first are the questions of justification. Are scientists really entitled to all the claims they make, or even to most of them? Are scientists entitled to say that any of their theories are actually correct? Scientists often make ambitious claims, but the history of science is a graveyard of ambitious claims now rejected. So philosophers of science ask whether scientific methods can be justified and, if so, what those methods can be taken to produce. Do they produce the truth about the world, accurate predictions, reliable technology, helpful mythology, or what? For each of the aims science might have, one can ask whether the methods scientists use are suitable. Are the means suited to the ends? Can it be shown that the methods are really going to deliver what it is claimed they should deliver? These are questions of justifying science.

The second kind of question philosophers of science ask sounds more modest than the questions of justification. This is the question of description, the simple request for a general description of what is going on in science. Here the point is not to show what science really achieves or to defend its methods, but just to manage a better understanding of how science works, of what its methods are, for better or for worse. If this is all one wanted to do, just to describe how scientists test their theories, say, then one might think the job would not be very difficult. Just ask a friendly scientist and she will tell you.

Of course it does not work that way. There is a great gap between what people can do and what they can describe. It is one thing to be very good at riding a bicycle, quite another to be very good at giving a general account of how a bicycle is ridden, of the physics and the physiology involved. A person can be very good at doing it and very bad at describing it. To take another example, it is one thing to be able to speak a language fluently and so to be able to distinguish grammatical from ungrammatical strings in the language, but quite a different thing to be able to describe the principles that guide that judgement. Science is no different. Scientists may be very good at doing what they do, but they are not very good at describing what they do. I do not say this out of a feeling of philosophical superiority. Philosophers are pretty awful at describing what scientists do as well: it is just a very hard problem. But it is one of the central problems of the philosophy of science.

These are some of the questions philosophers of science ask; but what is the point? This is a question often asked about philosophy generally. Why isn't it all just a waste of time? What is the good of it? One of the reasons people ask this question is because philosophy does not seem to have any associated technology – philosophy does not bake bread or build bridges. The philosophy of science, however, seems a possible exception, one of the few areas of philosophy where there might be a technology, broadly construed. Not bridges, but some people have hoped that the philosophy of science could make itself useful by helping scientists. That would be the practical application of the philosophy of science: it would make better scientists.

Unsurprisingly, many philosophers of science have been rather keen on this idea. Personally, however, I do not put much stock in it. The prospects of the philosophy of science providing extensive assistance to practising scientists nowadays are dim. Of course I do not conclude from this that the philosophy of science is a waste of time. An astronomer may devote his life to a better understanding of the stars without hoping to influence their behaviour. Similarly, a philosopher may hope to achieve a better understanding of how scientists work, without hoping to influence them. It may turn out to be useful as well as enjoyable for budding scientists to study some philosophy of science as undergraduates, but the justification of the discipline does not depend on this. Science is a central and pervasive part of our culture and our lives, and the attempt to understand better how it works and what it achieves is fascinating and worthwhile for its own sake.

THE CAMBRIDGE CONTRIBUTION

For the last forty years, there has been a thriving Department of History and Philosophy of Science at Cambridge, a Department that has become one of the outstanding centres for the history and philosophy of science in the world, with important work going on in many areas of the subject. Unparalleled library and archive resources and an extraordinary variety of research seminars attract scholars and graduate students from many countries. The Department also provides popular courses for undergraduate scientists, who can study history and philosophy of science alongside two other scientific subjects in their second year, and on its own in their final year. One of the secrets of our intellectual success is the productive way we combine the philosophy and the history, developing in tandem general accounts of how science works and particular accounts of how things actually went in particular places at particular times. We are also very fortunate to have one of the outstanding museums of the history of science and of scientific instrumentation in the world. Cambridge is a natural place to have such a wonderful resource, because Cambridge has been and continues to be the site of so much important and influential scientific research.

That is the last forty years in Cambridge for the history and philosophy of science – a world centre for the subject. But forty years is as nothing here; in forty years one is just beginning to figure out where the rooms are. To appreciate Cambridge's contribution to the philosophy of science, one has to go much further back. I propose to jump back not forty but four hundred years. This only takes us halfway, since the University is about eight hundred years old, but it already gives us an embarrassment of material, only a tiny part of which I can mention in this historical package holiday. Indeed I will only be able to touch on four Cambridge figures who have made important contributions to the philosophy of science; they will have to stand as representatives of many others.

Four hundred years takes us back to around the year 1600. The most famous figure in Cambridge philosophy of science at that time was Francis Bacon. He arrived at Trinity College at the age of twelve, and went on to become a lawyer, a politician, and one of the most influential philosophers of science of any period. Unsurprisingly, perhaps, his work in philosophy of science is more impressive than his work as a politician. Indeed as a politician he was corrupt, rightly accused of accepting bribes. He admitted as much, but offered the interesting defence that it should not be held against him since, although he took the money, he never let it influence him. This did not get him off, but he did not suffer too badly.

Whatever his political morality, Bacon was a gifted philosopher of science. He was a prolific and stylish writer, especially good at aphorisms such as the famous 'knowledge is power'. Science has two components which Bacon sometimes referred to as 'light' and 'fruit'. The light that science provides is insight into the inner workings of the world, in particular, the workings of its invisible parts. But Bacon emphasised that science is not just about light, it is also about fruit, about technology, control, and improving the quality of life. That scientific knowledge should have this power is an obvious thought for us, but it may not have been nearly so natural for people in Bacon's time. The standard image of science then seems not to be nearly so practical, and what Bacon had to say on this subject may have played a role in changing that image.

Bacon also emphasised the fact that science is not static, but changes and grows and, he held, would progress if scientists handled themselves correctly. Bacon was one of those philosophers of science who thought that philosophers could help scientists to do better science. He emphasised the importance of careful observation, and the importance of gathering data without prejudice. Observation, he claimed, should come before the hypothesis it is supposed to test, lest the scientist's attachment to her hypothesis bias her observation. Once a hypothesis has been formulated, scientists should look not for data that might support it, but for negative instances, for counter-examples.

This influential idea is nowadays associated with the work of Karl Popper. Suppose the hypothesis (I am afraid philosophers' toy examples are dull) is that all ravens are black. No matter how many black ravens are observed to date, it remains possible that there is one of another colour lurking around the corner. In other words, positive instances will never prove a general hypothesis. One non-black raven is however sufficient to refute the hypothesis, because even if all the other ravens are black it is still false to say that all ravens are black. This is a striking logical asymmetry: positive instances never prove, but negative instances disprove. Bacon made a great deal of this. According to him, if scientists want to make scientific progress they had better spend a lot of their research time eliminating hypotheses by finding negative instances, in the hope that the hypotheses that survive will be true.

Much of Bacon's methodological advice was negative in a different sense. He held, rather plausibly, that people are prone to what is now called 'systematic irrationality', and he set out to catalogue the various forms this irrational thinking could take, categorising them somewhat artificially under the headings of different 'idols' of the mind. Thus Bacon reported that people tend to be overly impressed by evidence that confirms their prejudices, that they are misled in various ways by language, that they focus on superficial features of the objects they study, and many other forms of cognitive disability. One interesting feature of the diverse idols of mind is how widely they apply to thinking generally and not just to science. This fits with Bacon's view that the methods of science do not differ fundamentally from other forms of enquiry, including the day-to-day thinking we must do to manage our lives. This is a view for which I have considerable sympathy, though many philosophers of science, including some we will come to shortly, have been concerned rather to emphasise the differences between scientific and everyday thinking.

Like most philosophers, Bacon was much better at talking abstractly about what he supposed science was like than he was at actually doing any science. It is not that Bacon was uninterested in experimentation, but some of his experiments were rather odd. Indeed there is a famous story according to which his death was due to one of his stranger experiments. The experiment consisted of stuffing a chicken with snow to see if this was a particularly effective way to preserve the chicken. The fate of the chicken is not recorded, but Bacon is supposed to have died from the influenza he contracted while conducting the experiment. In addition to his dubious experimental technique, Bacon turns out not to have been the most perceptive judge of the science of his time. He seems, for example, to have ignored the work of William

Harvey, the man who is credited with discovering the circulation of the blood, something we now regard as one of the major episodes in the history of science. What is strange is not just that Bacon gave no credit to Harvey, but that he appears not to have been aware of what Harvey was doing. This is particularly surprising since, as it happens, William Harvey was Bacon's personal physician.

If we now jump forward a century, we come to the Cambridge figure everyone has heard of: Isaac Newton. Another Trinity man, Newton is perhaps the greatest figure in the history of Cambridge and one of the greatest scientists that has ever lived. He gave us a unified account of the way things move, a beautiful theory of force and motion as it applies on earth and in space. His also did enormously important and influential work in optics, the study of the behaviour of light. Trinity College Chapel has a wonderful statue of Newton that is larger than life and towering overhead in a way that seems designed to encourage worship. In his hand, he holds a prism representing his research in optics.

Most scientists do not take much interest in the philosophy of science; they get on with their work without being particularly reflective or self-conscious about what they are doing. But some scientists do stand back from their own practice and attempt to understand better what they are doing, how the work should be done, and what it can be taken to achieve. Newton was such a scientist. He made important contributions to the philosophy of science. Indeed it appears that Newton's philosophy actually influenced his scientific practice; certainly he used philosophical arguments against his scientific opponents. This rather goes against what I suggested earlier about the general lack of influence of the philosophy of science on science, but then Newton was an exceptional scientist.

One philosophical dispute in Newton's time concerned the question of whether we should understand scientific theories as revealing hidden truths about the world, the realist view, or instead take them to be more like computers, whose purpose is calculation rather than description, the instrumentalist view. The point of theories on the instrumentalist view is not to describe a hidden reality, but to provide tools for calculating accurate predictions of the observable world. The dispute between realists and instrumentalists is a philosophical perennial and remains a central topic in the philosophy of science today. Newton took a clear stand on the dispute. He held that science should be in the business of uncovering real causes, the truth behind the appearances, and he is important in the philosophy of science partially because of the way he promoted this realist position.

Newton is also important in the philosophy of science because of the emphasis he placed on observation and experiment. Like the technological power of science, this is something we now take for granted, but it was not at all obvious to all the scientists and philosophers in Newton's time. He also realised the problem for science created by this dependence. If the reason for believing scientific theories is observation and experiment, then it seems that these theories can never be proven to be correct. The results of observation and experiment are never certain and, as we saw in the case of the black ravens, no amount of positive evidence will prove that a general theory is true.

Newton looked for a middle ground in his philosophy of science between conclusive proof or demonstration and mere conjecture. Demonstration, which many philosophers and scientists thought science ought to provide, is proof from self-evident first principles. Nice work if you can get it, but Newton realised that the role of observation and experiment rules it out. At the same time, he did not want a science that consisted of wild conjectures of merely plausible hypotheses. He sought a middle position, where although science does not generate the sort of proof pure mathematicians can provide, it is much more than guesswork. Newton claimed that somehow the data could be generalised to a theory that deserves high confidence even if it remained forever unprovable.

This helps to make sense of Newton's most famous philosophical slogan – *Hypotheses non fingo* – 'I frame no hypotheses' or 'I feign no hypotheses'. This seems a very odd thing for Newton to have said, since he spent his life framing a great number of wonderful hypotheses. Newton deployed that slogan in a particular context, defending himself against the charge that his theory of gravity was unacceptable because, while it used gravity to explain many other things, it did not properly explain gravity itself. Newton admitted that in that sense he did not attempt to explain gravity, but that such an explanation would be merely hypothetical and in any event not required in order to justify the physics he did provide. In more general terms, what Newton meant by his slogan, I think, was that Bacon was right, that science has to start with careful observation, and that the theory had to in some sense emerge as a warranted generalisation of the evidence. When Newton said that he did not frame hypotheses, what he meant was that he did not simply invent hypotheses that would account for the data, but rather found a path from the data to the hypotheses, even though that path could never be a path of proof. Whether there actually is any such general path from data to theory remains a central question in the philosophy of science.

To reach the third of my four Cambridge contributors, we now jump from around 1700 to around 1850, though we stay at Trinity. This contributor is William Whewell. He also has a nice statue in the College Chapel, as does Francis Bacon, though neither of them can hold a prism to Newton's overwhelming figure. Still, Whewell was an extraordinary polymath. He did seminal scientific work on the motion of the tides and he was at various times Professor of Mineralogy, Professor of Moral Philosophy, Master of Trinity, and Vice-Chancellor of the University. Whewell's great range of interests included questions of scientific terminology and he is, in fact, credited with coining the very word 'scientist'. What is striking is how late he did it, around 1840. Before then, people like Newton would not have been called 'scientists': they would have been called 'natural philosophers'.

Whewell was called the 'Professor of Moral Philosophy' not because he was particularly interested in ethics, but because his chair was in philosophy rather than in science.

One of Whewell's central interests was the history and philosophy of science (HPS), and one of the reasons that he is such an important figure in HPS is because he did both the 'H' and the 'P'. In the Cambridge HPS Department today, we work hard not to let the history of science and the philosophy of science become separate intellectual islands. It may seem surprising that effort should be required to avoid this, but sadly it is what happens in many HPS Departments elsewhere: the philosophers only talk to the philosophers, the historians only to the historians. The reasons for this are complicated, but one factor is the difference between the techniques of historical and philosophical investigation. Such a separation is however a terrific waste of intellectual potential, and our success here at Cambridge in bringing the two areas into productive interaction is a source of pride. Whewell is a model here, in the way he appreciated the importance of bringing the history and the philosophy of science together.

In his philosophical work, Whewell went against Newton, insisting on virtually the opposite of *Hypotheses non fingo*, at least as I have interpreted that slogan. Whewell claimed that scientists should be in the business of framing hypotheses in the sense that Newton proscribed: scientists should search for hypotheses that would unify the diverse evidence. Good evidence for a hypothesis is not just numerous and accurate, but also shows great *variety*, and that is one reason there is according to Whewell no simple path from that evidence to the hypothesis. Here Whewell emphasised an ancient idea about scientific understanding. On the surface the world is a mess, terribly complicated, because many different factors are interacting and we only see a small part of what is going on. Underneath the surface, however, we can find the fundamental forces, which may not be visible but which will reveal the unity and simplicity that underlies the superficial complexity. This ancient idea of unity beneath diversity has been enormously influential, and Whewell developed it in a particularly fruitful way.

The fourth of my four figures brings us into the twentieth century and moves us from Trinity to King's College. He is John Maynard Keynes, one of the most important and influential economists of the century. What is not so well known is that his first book, the book that got him his Fellowship at King's, was on probability. This work is of particular importance to the philosophy of science because of the way Keynes understood and interpreted the notion of probability.

The question of how we should understand claims about probability is a central philosophical topic. One view is that claims about probabilities are really claims about statistical patterns. Thus, to say that a coin has a probability of one-half on this view is just to say that, if you were to toss the coin many times, it would come up heads roughly half of the time. Keynes, however, argued that this is not the fundamental notion of probability, which is instead a relation between claims, between statements. More specifically, Keynes held that probability claims are claims about the support that evidence gives to a hypothesis where the evidence does not entail the hypothesis (Newton's worry), but where the evidence makes the hypothesis more or less probable. This sort of question about the relation between evidence and hypothesis lies right at the heart of the philosophy of science; hence the great philosophical interest of Keynes's work on probability.

We will return to Keynes, but first I want to consider what sort of common philosophical thread one might find running through these four Cambridge figures. In some ways what is more important is how they differed, intellectually and culturally. Certainly they had important philosophical disagreements, one of which we will shortly consider; but there is also an important common theme, the theme of empiricism. All four held that the fundamental source of knowledge about the world is observation. The contrast here is with rationalism, the view that it is fundamentally through thought, not observation, that we come to know how the world operates – René Descartes, for example, of *cogito ergo sum* fame, was a great seventeenth-century rationalist. The British, however, have tended to be empiricists, and our four figures run to form. They all emphasised that science cannot be done exclusively in the armchair; that scientists have to get out to do the experiments (though Newton did some of his best experimental work very near his armchair, in his rooms at Trinity). But they also realised the price scientists have to pay for taking the empirical route to knowledge.

Newton is explicit about the need to do science empirically, to rely on observation and experiment. In the *Optics*, he wrote that 'although the arguing from Experiments and Observations by Induction be no Demonstration of General Conclusions; yet it is the best way of arguing which the Nature of Things admits of'.[1] Scientists have to forego the certainty that proof provides because proofs are not to be had for claims about the way the world works. The only way to discover how it works is through observation. But this creates a difficult problem, at least from a philosophical point of view: the problem of gauging the uncertainty. Once the idea of proof is abandoned and replaced by the idea that the evidence supports or undermines a hypothesis, the relation is one of degree. It is not a question of 'Yes' or 'No', but of 'More' or 'Less'. This is one reason that Keynes's work on probability is so important: probability is a measure of the more and less. The difficult problem that philosophers have worked on is to understand the factors that determine this degree. What sort of evidence supports or undermines a theory, and what makes for more or less support? By considering some of the factors that philosophers have suggested, we can set the stage for the philosophical puzzle that I want to consider in the final part of this paper.

PREDICTION AND PREJUDICE

The factors that seem to increase the support for a scientific theory can be roughly divided into features of the evidence and features of the hypothesis or theory. On the evidence side, more supporting evidence is better than less. That is pretty obvious, but how much evidence the scientist has is not the only factor that affects support. Variety in the data is also an evidential virtue. A scientist who just repeats the same experiment over and over eventually reaches a point of diminishing returns, whereas a theory supported by a variety of experiments inspires greater confidence. Having accurate and precise supporting data is another evidential virtue, as is having the results of controlled experiments, where the scientist can be confident of the absence of disturbing influences. The same applies to so-called 'crucial' experiments, where the evidence simultaneously supports one theory while undermining some of its rivals, and to evidence that would be very improbable unless the theory was true.

One can construct a similar list of theoretical virtues. One is the prior plausibility of the theory: how natural it is and how well it fits with other claims the scientists already accept. Simplicity is another theoretical virtue: the simpler theory is often given a better chance of being correct. Other theoretical virtues include the plausibility of the auxiliary statements that have to be used to wring testable consequences out of the theory and the absence of plausible competing theories. These lists of evidential and theoretical virtues should make it clear both why the support a theory enjoys is a matter of degree and also why philosophers of science find it so challenging to account in detail for the impressive but not very reflective way scientists test and evaluate their theories.

This list of evidential and theoretical virtues is intended to be relatively uncontroversial, but I want now to focus on a disputed factor, a factor whose epistemic importance is the matter of much debate among philosophers of science. The dispute concerns the contrast between successful prediction and 'accommodation'. In a case of successful prediction, the scientist first has his theory and then goes on to deduce a claim about the outcome of an experiment or observation that has not yet occurred. He then makes the observation or performs the experiment and finds the predicted result. In a case of accommodation, by contrast, the scientist has the data in question *before* he constructs his theory, and proceeds to construct a theory around the data, ensuring that the theory he builds fits the data, accommodating the theory to the data.

The existence of this distinction between prediction and accommodation is granted by both sides of the debate: the issue is over its significance. The dispute is this: are predictions worth anything more than accommodations? In other words, should scientists give a theory more credit for its successful predictions than for its accommodations? Many theories will have both sorts of support to their credit: the theory will accommodate some data and predict others. The question is whether the predictions give a stronger reason to believe a theory than the accommodations.

In discussion, I sometimes try to settle the issue democratically, by having a vote. Members of the audience have three choices. First, they can vote for the claim that predictions tend to provide more support than accommodations; second, they can vote for the claim that the difference between prediction and accommodation makes no difference, it does not matter when the data are known; or third, they can abstain, if they have no clear intuitions on the matter. The results of such votes are fairly consistent and rather interesting. Most people do vote, and so presumably have a view on the issue. Of those who vote, most vote that prediction is better than accommodation, but a large minority choose the second option, that it makes no difference. So the issue is controversial, and not just among professional philosophers of science.

For evidence of this disagreement among the professionals, we need look no further than the Cambridge people whose philosophical contributions we have been celebrating. If you think that there is something special about prediction, that it does tend to provide stronger support, then you will find William Whewell on your side. He wrote that: 'It is a test of true theories, not only to account for but to predict, phenomena.'[2] That is as clear a statement as you could wish that prediction has some special value over what Whewell calls 'accounting' and what I have called 'accommodation'. On the other hand, if you think that the distinction between prediction and accommodation makes no difference, you are also in excellent company. In the view of John Maynard Keynes, 'the peculiar virtue of prediction or predesignation is altogether imaginary. The number of instances examined and the analogy between them are the essential points, and the question as to whether a particular hypothesis happens to be propounded before or after their examination is quite irrelevant.'[3]

Which then is the right answer? People who think predictions are worth more than accommodations often say that accommodations involve building a theory around the data, that this is *ad hoc*, and therefore provides little support for the theory. But this is not a good argument. What does '*ad hoc*' mean? It is Latin, so it sounds sophisticated, but all it literally means is 'purpose-built'. In this sense accommodation obviously is *ad hoc*: the whole point is to build a theory to fit the data. To say that it is *ad hoc* in its literal meaning is just to repeat that it is accommodation: it is not to say or to show that the theory is poorly supported or otherwise deficient. So on this reading, to argue that accommodating theories are *ad hoc* therefore they are poorly supported is to argue that accommodating theories are accommodating theories, therefore they are poorly supported, which is a *nonsequitur*, to use another Latin expression. On the other hand, the expression '*ad hoc* theory' is often used in English, at least by philosophers, in a derogatory sense that implies that the theory is poorly supported or otherwise unattractive. On that reading, the argument becomes that accommodating theories are poorly supported, therefore they are poorly supported. This is the opposite of a *nonsequitur* but equally flawed: it begs the question, assuming what was to be shown (in Latin, a *petitio principii*). Either way, the *ad hoc* argument fails.

Leaving the Latin behind, what other arguments are commonly given for the claim that predictions are better than accommodations? One is the argument from testing, according to which predictions are worth more than accommodations because it is only in its predictions that a scientific theory is tested, and it is only passing a test that gives a scientific theory genuine credit. The idea is that a test is something that could be failed, and it is only a prediction that a theory can fail. In that case, the theory is made to stick out its neck in advance and say how things will be, so that the scientist may go on to discover that things actually are not that way. So, if the theory passes this test, and things are found to be the way the theory said they would be, the theory deserves some credit. In accommodation, by contrast, the theory does not stick its neck out, it cannot be shown to be wrong, because the theory is constructed after the data and compatibility is guaranteed in advance.

An analogy helps to bring out the intuitive strength of the argument from testing.[4] Suppose that Jacob, my elder son, takes his trusty bow and arrow, shoots at a target on the side of a barn, and hits the bull's-eye. We are impressed and give him a lot of credit. Now Jonah, my younger son, steps up to a different barn, pulls back his bow and shoots his arrow. Then he walks up to the side of the barn and paints a bull's-eye around his arrow. We would give him rather less credit, for archery anyway. That is the idea behind the argument from testing. Accommodation is like drawing the bull's-eye afterwards, whereas in prediction the target is there in advance. This argument seems clearly to show why successful prediction should count more than accommodation.

Nevertheless, as it stands this too is a bad argument. It confuses the scientific theory with the scientist, the theory with the theorist. What is true is that only in the case of prediction does the scientist run the risk of getting egg on the face; it is only in the case of prediction that the scientist may have to admit to having made a false prediction. But we care about the theory here, not the scientist, and from the point of view of the theory the contrast between prediction and accommodation disappears. If the predicted data had been different, that theory would have been refuted or disconfirmed, but just the same goes for accommodated data. If those accommodated data been different, the theory that was built around it would also have been refuted. It is also true that, had the accommodated data been different, the scientist would have built a different theory, but that is not to the point. From the point of view of the theory, the situation is exactly symmetrical. So the argument from testing fails.

Perhaps then Keynes was right and the supposed advantage of prediction over accommodation is imaginary. Many philosophers of science agree with him. Nevertheless, I will end this chapter by suggesting where one might look for cogent arguments for the superiority of prediction. There are two promising types of argument. One is relatively straightforward; the other is a bit like trying to scratch the right ear with the left hand.

The relatively straightforward argument is the argument from choice. It depends on the fact that scientists can often choose their predictions in a way that they cannot choose which data to accommodate. When it comes to prediction, they can choose their shots, they can decide which predictions of the theory to check. Accommodated data, by contrast, is already there and scientists have to make what they can out of it. But how can this be used to show that predictions tend to provide stronger support that accommodations? A scientist will wish to make the strongest case to the scientific community that his theory is correct. So he has a motive for choosing predictions which, if correct, will give maximum support to his theory, not because they are predictions, but because they will exhibit the sort of evidential virtues mentioned before. Thus the scientist will choose predictions that allow for very precise observation, which would substantially increase the variety of data supporting the theory, and so on. Scientists will tend to choose predictions that will provide more support than in the case of accommodation, not directly because they are predictions, but indirectly because scientists have control over which predictions to check, control that is not available in the case of accommodation.

That is the straightforward argument from choice. It is probably cogent, but it does not show quite as much as one might hope. It shows why predictions *as a whole* tend to be more powerful than accommodations, but it does not give a reason for the more ambitious claim that a single, particular observation that was accommodated would have provided more support for the theory in question if it had been predicted instead. To try to make out that claim, we need a less straightforward argument, the fudging argument. It is related to both the *ad hoc* argument and the argument from testing we considered before, but it may avoid their weaknesses.

The fudging argument depends on an interesting feature of the lists of virtues, namely that some of the evidential virtues are in tension with some of the theoretical virtues. Here is an example. On the evidence side, scientists want the supporting evidence to be extensive and varied. On the theoretical side, they want the simplest theory. It is easy to have either one of these virtues on its own. If one just want lots of varied evidence one can just collect an encyclopedia full of facts; but the 'theory' that is their conjunction will be incredibly ugly because the facts are so heterogeneous. On the other hand, if all that matters is simplicity, that too is easy, so long as one doesn't mind about fitting any of the evidence. What is hard and what scientists want is simultaneously to satisfy both constraints. They want simple theories that nonetheless handle a great diversity of evidence.

Now for the fudging argument. When scientists have data to accommodate, they do the best they can. If the data are diverse, however, this can lead to a sacrifice in simplicity and other theoretical virtues. That is what I mean by 'fudging': the scientist may, perhaps subconsciously, fudge the theory, putting in a few epicycles or extra loops to ensure that more of the data gets captured. In a case of prediction, by contrast, the scientist has no motive to introduce anything unnatural into the theory, because she does not know the right answer in advance and so would not know what kink to introduce into the theory even if one were required. So in this case the scientist will use the simplest theory and, if the prediction is successful, will have exercised both empirical and theoretical virtue.

The advantage that the fudging explanation attributes to prediction, is a bit like the advantage of a double-blind experiment that a doctor might perform to test the efficacy of a new drug. In a double-blind experiment, neither the doctor nor the patient knows which patients are getting the placebo and which are getting the drug. The doctor's ignorance makes his judgement more reliable, since he does not know what the 'right' answer is supposed to be. The fudging argument makes an analogous suggestion about theoreticians. Not knowing the right answer in advance – the situation in prediction but not in accommodation – makes it less likely that the scientist will fudge the theory in a way that makes for a poor support. If you think about the puzzle of prediction and accommodation for yourself, as I hope you will, you may think of some objections to the fudging argument, but the argument may give one of the reasons why predictions can be more valuable in science than accommodations – one reason why, on this issue, Keynes was wrong and Whewell was right.

NOTES

1   *Optics* (1704), 404.
2   Whewell (1847), aphorism 39.
3   Keynes (1921), 337.
4   Cf. Nozick (1983), 109.

FURTHER READING AND REFERENCES

The philosophical work of the Cambridge figures mentioned in this chapter are discussed in Fisch and Schaffer (1991), Gower (1997), Losee, (1993), and Quinton (1980). Two good anthologies of recent work in the philosophy of science are Boyd, Gasper, and Trout'(1991), and Papineau (1996).

Boyd, Richard, Philip Gasper, and J. D. Trout (eds.). 1991. *The Philosophy of Science* (Cambridge, Mass.: MIT Press)

Fisch, Menachem, and Simon Schaffer (eds.). 1991. *William Whewell: A Composite Portrait* (Oxford: Oxford University Press)

Gower, Barry. 1997. *Scientific Method: An Historical and Philosophical Introduction* (London: Routledge)

Losee, John. 1993. *A Historical Introduction to the Philosophy of Science*, 3rd edn., (Oxford: Oxford University Press)

Keynes, John Maynard. [1921] 1973. *A Treatise on Probability* (London: Macmillan).

Newton, Isaac. 1952. *Optics* (1704) (New York: Dover Publications)

Nozick, Robert. 1983. 'Simplicity as fall-out', in L. S. Cauman, I. Levi, C. D. Parsons, and R. Schwartz (eds.) *How Many Questions?*, 105–19 (Indianapolis: Hackett)

Papineau, David (ed.). 1996. *The Philosophy of Science* (Oxford: Oxford University Press)

Quinton, Anthony. 1980. *Francis Bacon* (Oxford: Oxford University Press)

Whewell, William. 1847. *The Philosophy of the Inductive Sciences* 2nd edn., 2 vols. (London: J. W. Parker)

CHAPTER 8

European citizenship and education

MADELEINE ARNOT

'Citizenship means more than cricket, teachers warn' ran the headline in the *Times Higher Education Supplement*[1] when describing the findings of a research project based in Cambridge and funded by the European Commission (EC). Cricket in England has often been associated with citizenship. Indeed Norman Tebbit, when Minister for Trade and Industry in Mrs Thatcher's government, took the view that the ultimate test of national identity and 'Englishness' was the cricket match. Asian and Afro-Caribbean Britons who had lived all their lives in the United Kingdom, he argued, would be hard pressed to support the English cricket team against countries such as India or Pakistan or the West Indies. Their true loyalties of citizenship would indeed be revealed.

In this chapter I would like to explore English approaches to citizenship especially when challenged by the increasingly strong European agenda. Ever since Alexis de Tocqueville described what he called the 'peculiarities of the English' in the 1830s we have been engaged in discussing such peculiarities with a mixture of 'celebration' but also 'lament'.[2] Since the 1960s such renowned historians as E. P. Thompson, Perry Anderson, Eric Hobsbawn, have debated, for example, the peculiarities of our modern state. Why for example, was it that England was one of 'the last of the major nineteenth-century powers to create a national system of education' and 'it was also the most reluctant to put it under public control'.[3]

To the outside observer, the English state educational system must have appeared until very recently as one characterised by an extraordinary lack of central control, a wide diversity of institutions, and a 'chronic lack of integration' between its parts.[4] It was only in 1988 that a national compulsory curriculum for all primary and secondary schools was established, albeit superimposed upon different models of local provision, a wide range of school types (voluntary, state, church, single sex and coeducational, selective and non-selective) and enormously varied levels of academic performance between schools.

'Englishness' as far as the educational system was concerned came to be associated with voluntarism, local diversity, and devolved administration. Indeed if anything it is a system premised not upon the concept of the good of the nation, but rather on the need to meet the needs of particular groups in society. Schools seem to be meant to serve private interests of parents rather than the collective notion of society. The image of the ideal citizen was that of the English gentleman (playing cricket) – a prototype of what Professor Brian Simon[5] once called 'an imperfect citizenship' based as it was on the elite strata within the English school system. It was not an educational ideal normally associated with mass schooling. What characterised our system of education was the way it was designed to separate out different groups rather than encourage common experiences and a common culture. The paradox of the English, he argued, was that whilst formal democratic rights of citizenship were being extended to the population as a whole, the system of education was deliberately accentuating social or class differences.

The 'singular' English attitudes concerning the state, freedom, individualism and citizenship are now under scrutiny in a way that many of us in education would not have believed twenty years ago. Such fundamental concepts were barely challenged despite the radical educational debates on social equality and civil rights in the 1960s and 1970s. The emergence of something called 'Europe', however, which attempts to merge often sharply contrasting national traditions through a common economic and social agenda, requires us to consider afresh the nature of our political traditions, not least in relation to any future role our educational system might wish or need to have.

In this brief discussion, I consider how such debates are affecting our thinking not just about European citizenship and what that means for English education. I hope to show you what it means for a group of educationalists in the School of Education in Cambridge University who are grappling with this new political Colossus. Our work in its own way represents in its infancy and in its diversity, some of the opportunities for research and educational practice of the new Europe (indeed much of our research so far has been funded by the EC or an associated European agency). The learning curve has been steep for all of us. For many years now educational research in the United Kingdom has been criticised for being rather parochial and inward-looking. Now we are rapidly becoming that new breed of Euroacademics who head down to Heathrow to fly to yet another European city.

In different ways, our work also reflects the response of the English educational system to issues of democracy and citizenship, and to new concerns about the creation of European identities. Although not yet adequately integrated, our research themes provide some insight into the sociological, philosophical and educational questioning of our English liberal democratic traditions by the requirement that we should now be promoting a European identity in all our work. Such a goal could only ever have been controversial for a country which had a history of confusion and ambivalence about how to educate for democracy, citizenship, or national identity.

EUROPEAN CITIZENSHIP AND THE NEW
EDUCATIONAL DEBATE

Education only recently became part of mainstream European policy in the Treaty of Maastricht in 1992.[6] Yet by the early 1980s, Ministers of Education had begun to talk about developing a *European dimension* within each national educational system in the Community. Although not prepared to direct member states in an area of such sensitivity, the Ministers of Education called for a concerted effort to put this new dimension into the school curriculum, into teaching materials and approaches, and teacher education. 'Europe' was associated in their minds not merely with the formation of a new union of countries whose economies offered enormous opportunities in terms of work, and cultural collaboration, but with the promotion of democracy. Drawing on the Copenhagen Declaration of 1978, members of the community were asked, amongst other things to:

> strengthen in young people a sense of European identity and make clear to them the value of European civilisation and of the foundations on which the European peoples intend to base their development today, that is in particular *the safeguarding of the principles of democracy, social justice and respect for human rights* (Copenhagen Declaration, April, 1978) (my italics).

Young people were also to be encouraged to engage with the development of the Community, to become 'aware of the advantages which the Community represents, but also the challenges it involves, in opening up an enlarged economic and social area to them'.[7] 'Thinking European' by the 1990s meant becoming European.

Member states were asked to engage with three sets of ideals, which Michael Evans, based here in Cambridge, and his colleagues in other English universities[8] describe as: the Community ideal; the humanist ideal; and the international perspective on citizenship. In each case, England appeared to balk at the fence.

The Community ideal: 'entails a range of assimilative notions such as belonging, citizenship, identity and integration'. Whilst the Dutch and German government referred explicitly to preparing the citizen, the United Kingdom (along with Spain) talked about 'preparing young people to take part in the economic and social development of Europe'.

The humanist ideal: encouraged the values of 'peace, human rights, freedom, democracy and understanding'.[9] The United Kingdom referred to 'helping pupils acquire a view of Europe as a multicultural, multi-lingual community' whilst the Spanish response highlighted political and democratic structures, and the Germans talked about achieving 'self- determination in freedom'.

The international perspective: asked for multiculturalism, solidarity, and intercultural education. Whilst the United Kingdom wished to promote an 'understanding of the EC's interdependence with the rest of Europe and the rest of the world', Germany thought that Europeans should learn to appreciate other perspectives and to encourage tolerance.

Whilst the 1988 Resolution promoted the 'safeguarding of the principles of democracy, social justice and respect for human rights', the United Kingdom responses did not include any such mention.[10] Our minimalist interpretations of the European initiatives in education not surprisingly led to what has been called a 'proverbial patchwork quilt' in terms of United Kingdom school responses.[11]

The 1992 Treaty of Maastricht mainstreamed education, making it central to its political project. One commentator described the new European agenda as that of creating 'unity through diversity'[12] – a notion of a European identity over and above national cultures. The Community would now actively encourage exchanges between students, teachers, and academics through a range of funded projects; encouraging the teaching of languages and the sharing of information and technologies and the promotion of distance education. The concept of a European identity was becoming one of European citizenship – a feeling of belonging to a community, feeling committed to it and responsible for its development. Education for European citizenship would now be expected to involve pupils in 'thinking, feeling and doing' Europe.[13]

However the signing of these treaties by the United Kintdom was unlikely to be followed by a smooth process of transition towards the cultivation of an explicit European citizenship. The National Curriculum for all pupils in England, introduced in 1988, had only just been put in place, leaving little room for manoeuvre by teachers or time to fit a new European dimension into the school timetable. *Education for Citizenship* (one of the cross-curricular themes promoted after the National Curriculum was in place) had not received sufficient official support for it to be implemented effectively.[14] Teachers had not received guidelines and training to teach such a subject and certainly would feel ill prepared to give such a controversial topic a European slant.

Feeling European

Not surprisingly, given this political history, the level of critical political awareness about Europe amongst English pupils is very low. Last year a large-scale student survey of 1,337 fourteen- to sixteen-year-olds in England, France, Germany, Italy, the Netherlands, and Spain conducted by my colleague Michael Evans and others[15] found that:

1 Only 19 per cent of English teenagers replied 'yes totally' to the question do you consider yourself as European, compared with 90 per cent of the Dutch and 68 per cent of the Spanish.
2 Pupils' reasons for wanting to learn more about Europe were somewhat pragmatic:
   Help with travel and holidays (28 per cent)
   Understand social and economic problems better (18 per cent)
   Learn more languages (18 per cent)

Learn about other people's culture and way of life (17 per cent)

Improve job prospects (15 per cent).

Of those who were positive about Europe, comments such as these from Italy and another from France suggest a more developed political awareness than many of the English pupils:[16]

'As a citizen of Europe I'd like to be better informed so as to act in the best way possible.' (Italian)

'I would like to be well informed about Europe and Maastricht because later I will be a voter and for that it is necessary for me to have increased skills in order to vote properly.' (French)

The response of schools in the United Kingdom to the new European funding initiatives was characterised as *inspired opportunism*.[17] A number of schools took EC funds to develop such partnerships. As colleagues in Cambridge found when they evaluated the use of Community funds to promote the European dimension in a sample of English schools, teachers were struggling to define what was meant by a European dimension and what it would mean to engage in exchanges and partnerships with colleagues in other countries.[18] Initially subjects such as social studies, modern languages, the expressive and aesthetic arts, history, geography were relatively easy to use to promote knowledge about Europe but increasingly other subjects and other teachers were becoming involved.

Contradictory effects were reported for pupils involved in such school partnerships. Positive benefits included a reduction in stereotyping and anti-European feelings, improvements in language learning, a greater interest in Europe and a greater desire to communicate, more positive attitudes to living with families abroad, and increased awareness of differences between countries. Significantly however, few teachers reported that such projects had effected change on issues to do with equal opportunities, justice, immigration, defence and democracy.

Pupils also reported that these new projects had little effect on their attitudes to Europe; only 23 per cent felt more European as a result of being involved. Indeed only 28 per cent thought of themselves as European anyway. Again, a slightly larger percentage of girls than boys admit to a European identity with a larger proportion of girls saying they would consider living and working in another European Union (EU) country. Whilst 63 per cent of pupils thought it was a good thing that the United Kingdom was a member of the EU, 31 per cent reported being 'not bothered', despite the overall success in school terms of the collaborative schemes.

Without any systematic set of evaluation criteria, however, and little monitoring, such school projects were already generating inequalities. Anthony Adams, Michael Evans, and John Raffan in the School of Education found that the majority of pupils who had participated in such projects in forty-eight co-ordinating schools were female and white. Male pupil involvement was limited and that of children from minority ethnic communities was negligible.[19] These results were perhaps not surprising, given the high achievement of girls in modern languages. Is Europe becoming a female project? A fascinating thought.

Ethnic patterns of participation in the new European context are also to become matters of major concern in the future. Convery *et al.* (1977) found that only 5 per cent of Afro-Caribbean British and 10 per cent Asian-British pupils thought of themselves as European: in contrast with 48 per cent of white British. Certainly for these pupils (rather than the successful linguist) the concept of European identity, like a national identity, is highly problematic.[20]

Political engagement

Not surprisingly such exploratory research raised more questions than it answered, touching as it did on English pupils' relative lack of political engagement. Pupils' ignorance on European matters and structures, despite their expressed interest, must surely be one of the failings of our educational system. In the context of Europe, it is a particularly dangerous failing. As Andrew Phillips, founder of the Citizenship Foundation, recently pointed out, each year the European agencies sent out something like 6,000 regulations and directives. Pupils' lack of knowledge becomes critical at the time of greatest need for active participation in shaping the political agendas, especially given the decline of local democracy, the rise of a quango state and the bureaucratisation of power in the European Union in the 1980s and 1990s.[21]

Evidence recently presented by Professor Ivor Crewe (Vice-Chancellor at Essex University) provides an even more dismal confirmation of the very low engagement of the English.[22] Professor Putnam at Harvard University and Professor Crewe, having conducted 3,000 interviews with individuals in many different communities in the United Kingdom and in the United States, concluded that: 'Civil society is under stress' in both countries. The crisis, they argue, is between citizens and state: a crisis in relation to what they call *civic engagement* and *public discourse*. Both these features are essential to the effective running of democratic institutions of government, to the levels of engagement of individuals to their communities, and to the development of tolerance and the duty to defend the rights of others.

Although Crewe found that the British were more likely to associate citizenship with responsibilities towards a community and collective identities, in fact civic engagement in terms of civic association, community groups and voluntary work was very low. Up to a third of the British sample reported not being engaged in any civic activities at all (compared with only 6 per cent in the United States).

The results from the United Kingdom on public discourse were even more depressing. The researchers set out to discover how many individuals had had more than a five-minute discussion in the previous month on any of fourteen topics: for example, international affairs, EU domestic affairs, economy, local concerns, schools/traffic. Two-thirds of the English sample had recalled having discussed none (or only a few) of these issues only once or twice in the last month (compared with only 10 per cent who reported having had frequent discussions of these topics). Seventy per cent of the sample seemed to support the rather classic English stereotype that one should not talk about religion or politics. Apparently we are all watching television.

What is so disappointing is the evidence from both these studies that young adults in the United Kingdom seem not to have absorbed the practice of public discussion and political engagement. For some, such as the head of the Schools Curriculum and Assessment Authority, Nicholas Tate,[23] such evidence indicates that there has been a moral decline amongst young people. Britain in the post-war era is being criticised for having parents who are lax in teaching morals and standards of behaviour, teachers who are far too taken with relativist notions such as multiculturalism and young people (the baby boomers of the 1950s) who are far too dependent on the nanny welfare state. The solution, he argues, is to set up a new curriculum for social and moral education based on a set of core values (the new ten commandments). Such an education would awaken the 'good citizen' in the new generation.

Dr Tate does not explain the 'failure to belong' to what others describe as the 'obsessive individualism' of the last two decades, an individualism many now see as symbolised by Mrs Thatcher's now infamous comment, 'There is no such thing as society – only individuals.' He points instead to the failure of the English to develop a strong national identity and advises against any pre-emptive move to develop a European identity.

THE 'GOOD' BUT GENDERED CITIZEN

In my mind there is no better way to find out how we understand ourselves as a nation than to research in the European context. One of the greatest benefits of working with Europe is the opportunity it offers to reconsider the relationship of English culture to that of the Continent, and indeed to unravel the specificity of our educational and political traditions.

Comparative educationalists contrast English educational traditions with those on the Continent. I am sure that many of us could easily slip into the stereotypical images of what de Madariaga described as easy to use stereotypes of, for example, English *action*, French *thought*, and Spanish *passion*.[24] Martin McLean's description of the three great European epistemological traditions is more sophisticated. His insights suggest that knowledge and thought (epistemological traditions) have different functions in each national culture.

In a fascinating analysis, McLean describes the European traditions of encyclopaedism, humanism and naturalism. Encyclopaedism is based on the principles of universality, rationality and utility, where all students are made to learn as much as possible about all valid subjects, which are valuable not just for their own sake but for their use. This tradition is best represented in the French educational system. At the other end of the spectrum can be found Naturalism, which emphasises the development of individuals in the real world through the promotion of practical work, creativity, and learning through experience.

McLean suggests that humanistic traditions in education were found in England. Here the focus was upon morality, individualism and specialism. The Christian gentleman was to be educated in moral sensibility, commitment to duty, and the capacity to make informed decisions. A separation of academic from technical/vocational subjects characterised this tradition in the past. However this tradition has been severely tested in the twentieth century by an increasingly individualist ethos. Not only has the gentlemanly ideal of a Muscular Christian (athletics and Christianity) been replaced by new heroes – for example, the sensitive athlete, the glamorous TV producer – but the emphasis on vocational/industrial ethics in schools has removed many of the humanistic values of a broad and balanced education. I am sure that one can find similar tensions in the modernising of each of the other epistemological educational traditions.

I shall turn now to a set of more recent comparisons – those found in one of our Cambridge Gender and Citizenship projects.[25] Over the last three years I have been engaged with colleagues from Greece, Spain and Portugal in exploring with a group of student teachers (who, having just finished training, were about to become the new generation of teachers) how they understood men and women's position in contemporary society today, especially in relation to their concept of citizenship.

What we found, using surveys, interviews and single-sex focus group discussions, was that gender differences play a considerable part in shaping people's concept of citizenship, albeit in different ways in different countries. The differences in male and female pupils' attitudes to European identities, already discussed, were perhaps more significant than might be anticipated. Was it in fact the tip of the iceberg suggesting far more substantial gender differences in citizenship values? Is this why in so many European 'democracies' women play such a small part in determining the political life of their countries?

The gender angle is rather new to the citizenship debate, despite the fact that so many models of the ideal citizen are clearly gendered. As far as the European Union is concerned, women are still not participating in economic, cultural, and political decision-making in public life. Although women are performing exceptionally well in many European educational systems, many are not translating such success into highly skilled/ professional employment, nor are women making sufficient gains in the public sphere. The labour market in Europe is still deeply sex segregated, with women often marginalised and over-represented amongst those experiencing poverty.

The bottom tier of Europe consists of countries such as England which, until recently, had only 18.5 per cent female representation in its national assembly (raised recently by the election of a Labour government). The United Kingdom is in twelfth place out of forty European countries. As the Secretary-General recently commented: 'Without equality we cannot have fully representative government or social justice'[26] Women were still severely under-represented and the situation is actually worse now than four years ago.

Contemporary feminist political theory has much to tell us about how the long tradition of European political thought has presumed a difference between the sexes and based a distinction between public and private spheres on such differences. Women were therefore excluded from the concept of liberal democratic citizenship – a form of citizenship which, some argue, is based upon a 'fraternal contract'.[27] The question of how European civic traditions affect contemporary teachers is the central concern of our project. Learning about our civic traditions along the lines suggested by the European Ministers of Education in 1988 must not mean learning about a gendered, indeed male, concept of citizenship.

In our discussions with groups of student teachers in five countries, we found that there was not one way to talk about citizenship – there were a number of different sets of discourses, each of which constructed notions of the good citizen. Put simply, what makes us specifically European are the political philosophical traditions, the religious traditions, and the developments associated with the rise of national welfare states. What unites the groups of student teachers we studied are these common European discourses; what defines our differences are our political and economic histories (for example, ideologies by means of which the state has achieved legitimation and the role of social movements).

In 1995, we surveyed a total of 958 student teachers across five countries about their levels of knowledge about men's and women's positions in society and then ran discussion groups which focused on how this new generation of teachers conceptualised citizenship.

Like others before us, we found that the concept of citizenship was not part of everyday speech. Although an English word, the concept of citizenship was problematic for the student teachers in our sample. As three English student teachers commented:

'I think I'm a citizen but I mean I'd have to delve into a dictionary really.'

'Oh I am a citizen. That doesn't mean anything to me really.'

'I don't know. I think there is such a thing as citizenship.'

The group of student teachers, as in the other countries, struggled to define the geo-political community to which citizenship could apply. They ranged from references to different national clusterings, the United Kingdom, Great Britain, England, Scotland, Wales, and Northern Ireland, the international or the global community. Some thought citizenship was about being a member of a village, a region. For others citizenship only had relevance when going abroad, or meeting strangers.

Languages of citizenship

Three distinctive languages about citizenship (discourses) appeared to have shaped the ways these young professionals made sense of the word *citizenship* – political, moral, and egalitarian. Not one but several of these different discourses are drawn upon to make sense of contemporary citizenship, each of which have a gender dimension.

Discourses of Duty

These are Greco-Roman concepts of political life and the duties of the state in relation to civil society and to the individual, as well as the duties of the individual in relation to the state. Political discourses appear both to refer to men and to be used by men. Such political discourses construct the relationship between the citizen and the state in ways that excluded the private. The family is outside this realm, as are women. The model citizen is one who fights for the abstract principles of liberty and democracy within the institutions of government. Within such discourses women (even those such as Mrs Thatcher) have no place. There are still deep conflicts between power and female sexuality. Female public figures are talked about as 'honorary men', or as 'bimbos' (sexually attractive and reliant on male favours).[28]

Greek male student teachers offered a strong ideal, that of the *critical citizen* – with a duty to fight for democracy over and above the state. Similar themes were found amongst Portuguese men. The Spanish male student teachers we spoke to saw a society of transition, making them more distant from the process of political responsibility. Student teachers in England and Wales in contrast tended to be aware of the arbitrariness of the state – they were critical of the moral behaviour, intentions, and actions of politicians. Some of them put forward images of the *sceptical citizen* ever on the alert, but not active in promoting democracy.

Greek political philosophy or the French Revolution appeared not to have shaped English and Welsh student teachers' understandings of the concept of citizenship. Without a history of teaching education for citizenship, few would have had access to classical traditions (which are mainly taught in private schools). Some would argue that with no major period of disruption to democracy and with considerable ambiguity between the status of a subject and that of a citizen, we would be unlikely in England to see the need to articulate such a political discourse. Our data, like that contained in Crewe's study, shows that of the five countries it is the English student teacher who is less likely to talk about the concept of 'duty' of the individual to the principles of democracy.

Morality, caring, and common values
Using the vocabulary and metaphors of Judaeo-Christian philosophies, especially those to do with core ethical values, virtues, and social conformity, such moral discourses bring together the concepts of culture, community, and the common good. The focus here is more on shared values, especially those which relate to Judaeo-Christian traditions about good neighbourliness, moral behaviour and caring. The person at the heart of this discourse values solidarity rather than excessive individualism: the virtues of loyalty, honesty, and sensitivity to others. A good

citizen is essentially 'a good' person. Good people are thus good citizens.

Whilst male student teachers such as those in Spain talk about being good professionals, obeying the laws (Greece) or following moral and ethical principles (Portuguese), female student teachers in these countries put forward models of the 'mother as reformer': 'My grandmother is a good citizen.' Women are described, for example, as 'family builders', exercising citizenship in the family and in private life.

British student teachers, on the other hand, struggled with such moral debates. The 'caring citizen' appeared to be deeply problematic. On the one hand, charity work might imply that the state was not doing its job properly; on the other hand, moral values are thought to be arbitrary. 'Goodness', as one English student teacher commented, is 'exceptionally boring'; it is also associated with elite cultures. As one male student teacher commented:

> 'All the connotations to citizenship we have in society are very middle class . . . I'd be a good citizen and play cricket, that type of thing. It has no relevance to most people . . . the average man in the street.'

The class imagery of the good citizen was particularly striking. For Welsh student teachers, the image was that of male respectability – a dated image of the bowler-hatted (English?) gentleman of the 1950s.

> 'A middle-aged balding fellow with a nice garden and a semi-detached house.'

> 'The citizen is a necessarily nice man in a bowler hat who has . . . a job . . . in the city and then comes back to his nice semi-detached house with a wife and 2.5 kids.'

> 'The first thing that springs to my mind about a citizen is that I get an image of a man in a bowler hat and a suit.'

Male and female student teachers in the United Kingdom on the whole presented a sophisticated awareness of just how heterogeneous our society is. They spoke of the importance of recognising social diversity, differences in ethnic and class cultures and values, the importance of not imposing arbitary values through schooling even if they are egalitarian values. Parental values are also seen as part of the complex moral climate surrounding schooling.

Social rights and egalitarianism

Egalitarian discourses derive from the humanistic liberal democratic traditions and are mainly shaped by the concept of rights, of individual freedom from oppression, from poverty and violence. Since the Second World War a range of social entitlements (housing, education, etc.) are brought into play. Egalitarian discourses, as one might expect, are the most developed in the United Kingdom – student teachers referred to a range of rights – those of children, the aged, gays, women, people with disabilities, etc. The language of rights referred not only to the work place but also to our concept of marriage and family life. Women teachers in the United Kingdom referred to the need to negotiate equality.[29] Such concepts are far less developed and are reported to be less likely to be implemented in Greece, Spain and Portugal. Here the good citizen is the *protesting citizen* – it was up to women to fight for such rights. The struggle to achieve equal rights was particularly poignant in discussions about family life in these countries, where personal autonomy needed to be achieved over and above domestic violence and a strong concept of mothering.

The impressions we received from our small sample of English and Welsh student teachers was one of their using strong egalitarian, confused yet sophisticated moral discourses and limited political discourses. European research allowed us to consider this in more depth and to begin to uncover the different ways in which the concept of citizenship is gendered.

In all countries the public sphere was associated with men and references to the political duties of the citizen were mainly found amongst men. Whilst the image of men in public life in all countries, including England and Wales, was *powerful*, and *competi-*

tive, the image of women in the public sphere was not strong; words chosen to describe women in the United Kingdom were *conscientious*, *efficient* and *competent*. We found that women here were thought to have only a marginal influence over public policy-making even though successful women abound in public life today.

Our survey also found that whilst in Spain, Portugal and Greece women were described in private life as *maternal*, in the United Kingdom over 40 per cent of our sample of student teachers chose to describe their image of women in domestic life as *efficient* and between 30 per cent and 40 per cent chose *competent*. Women were perceived in the United Kingdom to be largely in control of a range of private decisions (children), but also shared some responsibilities with men (e.g. childcare). Noticeably in the United Kingdom women were strongly represented as victims of sex-related violence.

In the case of men, no consensus could be found about the image of men in private life. In the United Kingdom less than four out of ten student teachers could agree on the descriptions *hesitant* and *disorganised*. One does wonder what British men are doing in private and domestic life to justify such negative images or such invisibility. At least in Portugal and Greece men were described as *paternal* in one of these roles.

EDUCATION FOR CITIZENSHIP

The marking of male public and female private spheres by our group of student teachers raises fascinating academic questions about the gender assumptions which underlie liberal democracy. The Gender and Citizenship projects raise practical political questions about how we should educate all pupils in contemporary society to become active, participating members of the polity. The seeming irrelevance of private life for issues of citizenship, for example, asks serious questions about how we are to prepare young people for current controversies about reproductive rights, such as the right to childbirth, *in vitro* fertilisation, abortion, surrogate parenting. Violence and abuse within personal relations and the family are clearly also matters of citizenship and are of vital importance to both women and men.

Our various projects in Cambridge also question how we should train teachers to teach issues of British and European or global citizenship. The failure particularly in the last decade to train teachers more effectively in issues relating to citizenship, social justice and human rights, and to methodologies which are sensitive to pluralism and diversity, is described in a recent book by my colleagues Anthony Adams (an English educator) and Witold Tulasiewicz (a comparative educationalist) as a 'crisis in teacher education'.[30] Competency-based models of training which reduce the higher educational component of professional socialisation offer little encouragement to those committed to European integration, multiculturalism, and social justice as educational ideals. Teachers' deskilling could well be contributing to the increase in social exclusion and alienation of groups of students from society.

Historically, civic education for pupils in England was invariably seen as 'high risk, low gain'.[31] In every decade attempts, it seems, have been made to introduce civic or citizenship education into our school. The model of citizen offered by school texts, particularly after the war has been dominated by the need to civilise the male workers, soldier and citizen. Lord Nelson's reputed battle cry before the Battle of Trafalgar could serve as the motto for the history of citizenship education – 'England expects every man to do his duty'.[32] A minimal number of alternative versions of citizenship have been available, which included women.

Civics versus citizenship

For the most part we have found in our research that the emphasis in England has upon been the teaching of *civics*, which can be identified with what my colleague Terry McLaughlin (a philosopher of education here in Cambridge) calls the *minimal* definition of education for citizenship.[33] This model involves mere transmission of information about civic matters and the development of basic values such as good neighbourliness. Here the emphasis is on conformity to core values, and limited involvement in community-based activities that are often local in character. We have rarely seen in the United Kingdom what McLaughlin calls the *maximal* version of education for citizenship, which is 'a richer thing' – involving the development of a much more extended and critical conception of what citizenship involves. Here there is insistence upon reflective understanding of civic and political concepts and issues and a much more explicit and wide-ranging commitment to democratic values and responsibilities, including an awareness of the requirements of the common good (for example, justice).

McLaughlin argues that such a conception of citizenship is compatible with a form of civic nationalism which is rational, flexible, pluralistic, and morally rich, in contrast with ethnic nationalism,[34] which is tempted by irrationalism, fanaticism and authoritarianism. A difficulty of this general approach is achieving a balance between the defensible requirement of criticism and critique in the task of political education involved in education for citizenship and the need to secure conditions of contra-individualistic identification and commitment. How can one secure solidarity in the context of a highly pluralistic multicultural society?

To some extent McLaughlin argues,[35] that education cannot avoid shaping a particular identity as well as a general or universal one. He quotes De Miastre's remark that

> I have seen in my times, Frenchmen, Italians and Russians . . . but as for Man, I declare I have never met him in my life.

A local identity is context bound, and, to some extent, the sense of being located is a prerequisite for freedom. This view is also taken by the new so-called communitarian movement (much of which is currently based in Cambridge), which suggests that we all need to be affiliated to a 'larger moral ecology' that embodies a social ethos, a consensus on the common good, and notions of loyalty and responsibility to the community as a whole. This *belongingness* gives a sense of coherence to our lives. It overrides narrow self-interest and the excessive individualism which can be associated with liberal democracy. It is from this general spring that many of New Labour's ideas appear to come.

But this line of argument gives rise to the perception that a European identity lacks the requisite features to function as a focus of this kind. It is difficult to avoid the potentially negative consequences of celebrating uncritically what has come to be called 'Fortress Europe' – a concept of Europe that is charac-terised more by its exclusions than inclusions; more by the rise in ethnic conflicts that its resolution of them.[36] Europe as a concept defined in Brussels still tends to be economic and strategic rather than cultural in character and therefore gives rise to a 'thin' rather than 'thick' form of identification and involvement. As McLaughlin points out,[37] the 'claims and concepts' of the Com-mission of the European Communities used in relation to the task of developing a European dimension are open to potentially objectionable interpretations in schools, and suggest a controver-sial role for higher educational institutions.

Should we therefore abandon the project of teaching a Eur-opean citizenship through schooling? Or should we start revita-lising those common roots which we found in our research on student teachers? Should we develop a new (less gendered) English national identity? Should we grasp the nettle and chal-lenge the peculiarities of our liberal democratic traditions, the individualism which lies at the heart of our system of education?

A new agenda

They say that the European community is only at an adolescent stage.[38] The research I have described here that is based in Cam-bridge could be said to be at an even earlier stage – it is embryonic. Many more issues are coming onto the agenda, not just the shape of education for democratic citizenship in Europe – but also the roles of mother-tongue teaching,[39] language awareness,[40] and teacher education[41] that are deeply affected by and affect current debates about European citizenship.

Such research offers ways of making a contribution to one of the most vital sets of questions about the role of our educational system in transmitting knowledge and shaping social identities in the next millennium. These issues are rapidly coming to the forefront in education and will increasingly, I believe, come to shape our educational research at Cambridge.

—

ACKNOWLEDGEMENTS

I would like to thank my colleagues Michael Evans and Terry McLaughlin for the comments on the first draft of this chapter and Tony Adams and Witold Tulasiewicz for their help in providing me with relevant material. I alone am responsible for any errors in the paper. My research is supported by a research fellowship from The Leverhulme Trust for which I am most grateful.

NOTES

1 *Times Higher Education Supplement,* 31 January 1997.
2 Green (1992), p. 213.
3 Ibid. p. 208.
4 Ibid. p. 208.
5 Simon (1994).
6 Lindlay (1996).
7 The Copenhagen Declaration, 1978, quoted in Convery *et al.* (1997), p. 5.
8 Convery *et al.* (1997).

9 Ibid. p. 6.
10 Ibid. p. 9.
11 Findlay (1996).
12 Sultana (1995).
13 Convery *et al.* (1997), p. 11.
14 See Whitty, Rowe, and Aggleton (1994) and Beck (1996) for a description of the situation in the United Kingdom and Edwards *et al.* (1994) for practice in other European countries.
15 Convery *et al.* (1997).
16 Ibid. p. 24.
17 Findlay (1996).
18 Adams *et al.* (1997).
19 Ibid. p. 2.
20 Convery *et al.* (1997).
21 Andrew Phillips's talk 'Citizenship education in the new context', Citizenship Foundation meeting at the House of Lords, 28 January 1997.
22 Crewe (1996).
23 Tate (1997).
24 Quoted in McLean (1996), p. 37; footnote 23 McLean (1995).
25 The first project on citizenship which focused on initial teacher training was funded by the EC; the second on gender and citizenship was funded by a Leverhulme Research Fellowship.
26 Quoted in *The Guardian*, 23 October 1997.
27 For discussion of feminist democratic theory see C. Pateman (1980); Jo-Anne Dillabough, Gabrielle Ivinson, and I have attempted to explore the implications of this feminist political theory for education, teachers' political identities, and representatives of public and private life. See Arnot (1997); Arnot *et al.* (1997); Arnot and Dillabough (1997); Ivinson and Arnot (1999, forthcoming).
28 Ivinson and Arnot (1999, forthcoming) and Arnot *et al.* (1997).
29 Ibid.
30 Adams and Tulasiewicz (1995).
31 Lawton (in interview 1995).
32 Brindle and Arnot (1998, forthcoming).
33 McLaughlin (1996).
34 Cf. Ignatieff, quoted in McLaughlin (1977), p. 62.

35 McLaughlin (1997), p. 9.
36 Coulby and Jones (1995).
37 McLaughlin (1998 forthcoming).
38 Convery *et al.* (1996).
39 Tulasiewicz and Adams (1997) explore the various approaches towards the teaching of the mother tongue in countries across Europe.
40 Work is being conducted in the School of Education by A. Adams on the exploratory field of language awareness with some EU funding and the help of colleagues in other European universities. This work brings together mother-tongue and modern foreign-language teaching and has the potential to allow children to use their experience of language to develop a critical analysis of their society and the fostering of intercultural relations (Tulasiewicz, 1998, forthcoming).
41 See for example Osler *et al.* (1995); Osler and Starkey (1996).

REFERENCES

Adams, A., M. Evans, and J. Raffan 1997. 'A pilot study evaluation of UK coordinated Socrates: Comenius Action 1 school-based partnership programmes', Research Report, Cambridge University School of Education

Adams, A. and W. Tulasiewicz. 1995. *The Crisis in Teacher Education. A European Concern?* (London: Falmer Press)

Arnot, M. 1997. 'Gendered citizenry: new feminist perspectives on education and citizenship' *British Education Research Journal*, 23, 3, 275–95

Arnot, M., H. Aráujo, G. Ivinson, and A. Tome. 1997. 'Changing femininity, changing concepts of citizenship: social representations of public and private spheres in a European Context', paper presented at the 3rd European Feminist Research Conference, Coimbra, Portugal

Arnot, M., H. Aráujo, G. Rowe, and A. Tome. 1996. 'Teachers, Gender and Discourses of Citizenship', *International Studies in Sociology of Education*, 6, 1, 3–5

Arnot, M. and J. A. Dillabough. 1997. 'Feminist politics and democratic values in education', paper presented at the American Educational Research Association, Chicago

Beck, J. 1996. 'Citizenship education: problems and possibilities', *Curriculum Studies* 3, 349–66

Brindle, P. and M. Arnot. 1998, forthcoming. ' "England expects every man to do his duty": the gendering of citizenship textbooks, 1940–66', *Oxford Review of Education, Special Issue on Political Education*

Convery, A., M. Evans, S. Green, E. Macaro, and J. Mellor. 1997. *Pupils' Perception of Europe: Identity and Education* (London: Cassell)

Coulby, D. and C. Jones. 1995. *Postmodernity and European Education Systems* (London: Trentham Books)

Crewe, I. 1996. 'Citizenship and civic education', paper presented at the Royal Society of Arts, published by the Citizenship Foundation

Edwards, L., P. Munn, and K. Fogelman (eds.). 1994. *Education for Democratic Citizenship in Europe – New Challenges for Secondary Education* (Amsterdam/LISSE: Swets and Zeitlinger)

Findlay, I. 1996. 'European dimensions in schools', in J. Ahier, B. Cosin, and M. Hales (eds.), *Diversity and Change: Educational Policy and Selection* (London: Routledge)

Green, A. 1992. *Education and State Formation: The Rise of Educational Systems in England, France and the USA* (London: MacMillan)

Ivinson, G. and M. Arnot. 1999, forthcoming. 'Public and private citizens: perspectives on gender relations and change' (1997), paper presented at the Gender and Education Conference, University of Warwick, to be published in M. Arnot and J. A. Dillabough (eds.), *Education and Citizenship: international perspectives* (London: Routledge)

McLaughlin, T. 1996. 'Citizenship, diversity and education: a philosophical perspective', *Journal of Moral Education*, 21, 3, 235–49

1996. 'Educating responsible citizens' in H. Tam (ed.), *Punishment, Excuses and Moral Development* (Aldershot: Avebury)

1997. 'National identity and education' in M. A. Santos Rego (ed.), *Educational Policy in the European Union after Maastricht* (Santiago di Compostela, Spain)

1998. 'The European dimension of higher education: neglected claims and concepts' in F. Crawley, P. Smeyers, and P. Standish (eds.),

Nations and Cultures in European Higher Education (Oxford and Providence, Rhode Island: Berghahn Books)

McLean, M. 1995. *Educational Traditions Compared: Content, Teaching and Learning in Industrialised Countries* (London: David Fulton)

—— 1996. 'School knowledge traditions' in J. Ahier, B. Cosin, and M. Hales (eds.), *Diversity and Change: Educational Policy and Selection* (London: Routledge)

Osler, A. and H. Starkey. 1996. *Teacher Education and Human Rights* (London: David Fulton)

Osler, A., H. F. Rathenow, and H. Starkey (eds.). 1994. *Teaching for Citizenship in Europe* (London: Trentham Books)

Pateman, C. 1980. *The Disorder of Women* (Cambridge: Polity Press)

Simon, B. 1994. 'Education and citizenship in England', in *State and Educational Change* (London: Lawrence and Wishart)

Sultana, R. G. 1995. 'A uniting Europe: a dividing education? Supranationalism, Eurocentrism and the curriculum', *International Studies in the Sociology of Education*, 5, 3–23

Tate, N. 1997. 'The European dimension in the school curriculum', presented at the European Parliament Office, 31 January

Tulasiewicz, W. 1998, forthcoming. 'Language awareness: a new literacy dimension in school language education', *Journal of Teacher Development*

Tulasiewicz, W. and T. Adams. 1997. *Teaching the Mother Tongue in a Multilingual Europe* (London: Cassell)

Whitty, G., G. Rowe, and P. Aggleton. 1994. 'Discourses in cross-curricular contexts: limits to empowerment', *International Studies in Sociology of Education*, 4, 25–42

CHAPTER 9

The University Botanic Garden

JOHN PARKER

In the Classical and Medieval worlds collections of plants were held as the only source of the drugs which were used in medicine. Thus for those who were studying medicine, and subsequently practising, a detailed knowledge of plants was essential. We still today obtain more than 50 per cent of drugs directly from plants: chemical synthesis, although a powerful approach to drug production, is expensive and limited in its application. Plants are remarkable natural chemical factories and this has been exploited by mankind throughout his evolution.

In Renaissance Italy in the late sixteenth century, the new universities began to gather plant collections in order to have material on hand to teach students about the medicinal uses of plants. These botanic gardens, essentially physic gardens, were developed in the universities of the city states of Northern Italy – Padua, Lucca, Florence and Siena. Physic gardens spread north of the Alps to Leiden in the Low Countries, and then to Oxford in 1621. Cambridge had a strong tradition in medicine dating back to the Middle Ages and by the seventeenth century there was a clear necessity for a botanic garden for displaying and teaching about the plants which were used for the training of physicians.

Although the need for a botanic garden was discussed in Cambridge University in the middle of the seventeenth century, it was only in 1762 that the University eventually managed to assemble sufficient funds and a sufficient consensus to purchase a private garden in the centre of Cambridge for this purpose. This garden was located on land which is now the New Museums Site, accessible from Downing Street and bounded by Free School Lane to the south and Corn Exchange Street to the north. Mr Mortlock's garden was converted to hold collections of medicinal plants in beds, with glasshouses for tender plants and lecture-rooms for students. Ornamental wrought-iron gates led from Downing Street into the Botanic Garden. The garden held a collection of about four thousand different plants which were used for teaching the basic botany underpinning the medical pharmacology needed by students.

The Botanic Garden set up in Mr Mortlock's garden survived for only eighty years. Two reasons account for this decline of the plant collection. Cambridge during the late eighteenth and early nineteenth centuries was growing rapidly. The University was expanding and new buildings were being constructed. The Botanic Garden, on its prime central site, was becoming over-shadowed. It became increasingly difficult to grow plants in this drastically modified environment. A further problem was described by S. M. Walters in *The Shaping of Cambridge Botany*, published by the Friends of Cambridge University Botanic Garden in 1981. Each plant in the Botanic Garden was labelled. The labels used were wooden and about one foot in length. These labels proved irresistible to the plagues of jackdaws which infested Cambridge. They were used as nesting material in this otherwise barren part of the city. When a blocked chimney stack on Pembroke College was taken down, about two hundred and fifty Botanic Garden labels were recovered from the nests of generations of jackdaws. Apparently, it was such a desperate situation that even when the gardeners went in for tea, the jackdaws descended and ripped out the labels. These dual problems were solved dramatically by the advent of one man.

This man, John Stevens Henslow, has had a formative influence on western European scientific thought, not directly through his own thoughts and writings, but through his influence on others. Henslow was appointed at a young age as Professor of Mineralogy and then six years later also took up the Professorship of Botany, thus holding two Chairs in the University. He had a knowledge of the natural world from the rocks themselves to the living organisms, and he was a great field naturalist. Henslow's influence on western thought is profound, because he introduced to this University the study of those aspects of biology which we would now regard as natural history. His Cambridge house was open to undergraduates whom he would welcome to talk generally about biology and geology, and he also led tours around Cambridge and district with the undergraduates, pointing out the things of interest as they walked. He drew remarkably able people to him, and the most able was Charles Darwin. Henslow redirected Darwin's thoughts away from theology and medicine and into natural history. Darwin's subsequent works are imbued with Henslow's eclectic approach – an integration of geology, soil science, botany, zoology and so on.

Professor Henslow realised that the University needed a new botanic garden, but that it had to be of a different type from the physic gardens typical of the world's universities. He concluded that botanic gardens existed for a totally different function not associated with medicine. They existed to provide collections of species which could be used for experimental science and 'vegetable physiology'. He persuaded the University to remove its two hectare garden to a fresh site, and to purchase eighteen hectares in order to put together a collection of the exciting species which were then being discovered all around the temperate world. These were plants from western North America and from eastern Asia in the temperate parts of China, the most spectacular being the trees. Thus the big trees and giant redwoods – *Sequoia* and *Sequoiadendron* – eventually found a home in the new Botanic Garden. He wanted space to grow trees for experimental purposes, not for medicine, and he wrote very compellingly that this

JOHN PARKER

was the sole reason for holding collections. Collections are held to enable research to be carried out to further the cause of science, or the cause of art. This was a great contribution and a major change in direction. He persuaded the University to purchase a green-field site south of the town in the direction of Trumpington Road in 1831. As we have seen, the University moves slowly, so that the first tree of the new Botanic Garden was planted in 1846, a mere fifteen years after the purchase of the meadow site.

An original plan of the Botanic Garden exists, which was drawn up by Andrew Murray, the first curator of the new site, between 1835 and 1840. In his job application, Murray had to produce an entire plan of the eighteen-hectare garden between Trumpington Road on the west and Hills Road on the east. Down the Trum-pington Road edge of the garden flows Hobson's Conduit, the artificial watercourse which brings fresh water from the springs at Great Shelford to the corner of Lensfield Road, and this was tapped as a water supply for the lake. The main elements of the Botanic Garden drawn by Murray exist to this day: a main east–west ride in a central position, the horseshoe-shaped lake to the north of the ride, the systematic beds to the south, and a range of glasshouses, not in their original position, the whole surrounded by dense groves of trees. Lack of funds immediately prevented exploitation of the whole area, and a scaled-down version was developed which left the eastern portion as allotments. It was only subsequent to the Second World War, when money became available from the bequest of Reginald Cory, that the whole of the University's holding could be brought into service as a Botanic Garden.

From the inception of the new Botanic Garden it was regarded as an important amenity for the University as well as a scientific resource. The quality of horticulture has therefore always been of the highest. This is exemplified by the work of the curator Richard Lynch, who was responsible, among other things, for building the glasshouses and developing the collection of bamboos on the northern edge of the lake. Lynch's plan of the Botanic

172

Garden is held in the University archives. It is similar to Murray's but occupying half the site, with the glasshouses at the north-eastern boundary.

The Garden as it exists today is essentially divided into two parts by a magnificent group of *Pinus nigra* and *P. sylvestris*. These trees occupy the boundary between the older garden focused on Trumpington Road, now 150 years old, and the fifty-year-old garden stretching eastwards to Hills Road.

The garden we see in the late 1990s owes a great deal to the generous donations of Reginald Cory, a scholar of St John's College who was fanatically interested in horticulture. He was brought up at Dyffryn House near Cardiff, surrounded by fine gardens. He poured his money and his love into the University Botanic Garden and he was able to change the nature of the Botanic Garden is a number of ways. His gifts enabled a house to be built within the Botanic Garden in 1926, known as Cory Lodge, as a residence for the first of the academic Directors, Humphrey Gilbert Carter (1921–50). The first three of the five Directors lived in Cory Lodge, but the house is now used as the Botanic Garden administrative offices and also contains the library. The book collection is a great treasure of beauty and scientific worth. It has both historical and contemporary value.

The Botanic Garden is a jewel composed of many facets which has been moulded by the brilliance of its horticultural staff. The garden houses a scientific collection of about ten thousand species for teaching and research laid out with such remarkable skill that they blend into an extremely beautiful garden. The different elements encompass the formal and the informal, in a natural and comfortable juxtaposition giving a garden of human dimensions – peaceful, sympathetic and reassuring.

The main gates are those of the original Downing Street entry to the first Botanic Garden of 1762 which were moved to their Trumpington Road location at the end of the main ride in 1909. The view up this main ride reinforces Henslow's vision of trees as the basis of our collection. Magnificent conifers march on either hand – *Sequioadendron* from California, cedars from Lebanon and Morocco, *Pinus nigra* subspecies from across Europe. The Cambridge tree collection is one of the most splendid in eastern England, now majestic and dignified in its 150-year-old power, setting the framework for the whole. As the focal point of this ride there is a fountain reflecting the formality and clean lines of Scandinavian-influenced design of the 1960s. It was designed by David Mellor, a steel artist from Sheffield who was working at the time at Robinson College designing candlesticks and light fittings. The fountain illustrates the essential pattern of growth of the Botanic Garden – each new element integrating with and enhancing the whole.

Cambridge is not noted for its streams. The chalky substrate sucks the water deep into aquifers. But the Botanic Garden is in the happy position that it can tap Hobson's Conduit to create a gentle brook bubbling over a puddled clay base to prevent loss of water. The stream gives an added habitat in the garden where beautiful and creative plant assemblages are displayed which change dramatically with the season. The stream feeds a lake, again an unusual landscape feature in the Cambridge region constructed, as were our canal systems, with a thick puddled clay lining. Around the lake, dense stands of vegetation have built up against a skyline of magnificent trees, giving a feeling of remoteness from the heart of Cambridge, only ten minutes distant.

Around the lake there is a woodland garden, a further unusual feature of the University Botanic Garden. Woodland gardens require an acid soil, about pH 6–6.5, whereas the soil in Cambridge is highly calcareous, with a pH of 7.8. The constant addition of leaf material to the soil over many years has enabled a remarkable diversity of woodland herbaceous plants to be built up sheltering under a diverse and beautiful tree canopy. In woodlands the habitat changes during the year: the woodland floor is shaded most of the year so many herbaceous plants flower in early spring before leaf emergence. The woodland environment is also moist and still, leading to a verdant, massive density of plant growth.

A completely different opportunity for growing and displaying plants is afforded by rock gardens. The major rock garden has been developed on blocks from a limestone pavement brought from northern England in the 1950s and supports a wonderful array of plants in an arid and exposed environment. The plants are arranged geographically encircling one arm of the lake so that we can compare and contrast the spectrum of species and their adaptations from the mountain regions of the world – North America, Europe, Asia and Australasia.

The Botanic Garden has only one major open space with fine-cut grass: the area backed by the Glasshouse Range along the northern boundary of the public area. This mown area is set off by beds of colourful Mediterranean plants, tender annuals and perennials, while the eastern border has a magnificent collection of slow-growing conifers in a *mélange* of shapes, colours and textures. Glasshouses enhanced the opportunities for horticulture. Three thousand frost-sensitive plants are housed within the complex and individual houses range from the homely to the highly exotic. There is a conservatory which could be developed at home in a cool glasshouse. This house is re-designed by one of the student gardeners every four months. Each student can therefore exercise his or her own talents in design, using their own choice of species and cultivars. The familiarity of the plants used in the conservatory makes it a favourite with visitors.

Research and teaching collections can also be grown under glass. A house is devoted to alpines and spring bulbs which have particular physiological requirements for their culture as a result of their patterns of growth and development in the extreme habitats where they are native. It is particularly important not to neglect the Tropics, where most plant and animal diversity is found. In the steamy hot tropical house the display of crops is particularly fascinating – banana, coffee, sugar cane, rice and yams spring up together, providing excellent material for teaching. In March and April crowds flock to the tropical house to see the superb jade vine. This liane from the Philippines has massive hanging racemes of extraordinary blue-green pea flowers. The colour is unique in plants and quite startling. In nature the flowers of the jade vine are pollinated by large bats who seek the abundant nectar. In our glasshouse the flowers are overrun by ants.

The systematic beds are of a construction which is unique in the world's botanic gardens. They were shown in their current arrangement on Andrew Murray's original map. The beds represent an artistic installation or construction since they are a two-dimensional array of species based on a particular taxonomic philosophy. The beds follow the taxonomic classification of Alphonse de Candolle who published a book in 1819 which was a major comprehensive account of flowering plant families and their definition. Each bed, either large or small, contains only one family but some of the large families, such as Compositae, occupy many beds. What Andrew Murray designed in 1835 was a physical array which reflects the book: the bed containing Ranunculaceae represents page one of the book and as you walk the outside belt the families occur in page order. De Candolle was one of the first to distinguish between what we call the monocotyledons and the dicotyledons: the monocots have single seed leaves (grasses and orchids for example) while the dicots have two seed leaves (such as buttercups and dandelions). The centre circle contains the monocot families and the surrounding torus the dicots. The whole installation consists of about 1,600 species densely planted in irregularly shaped island beds and it is clear that this whole garden feature was developed with both aesthetics and science in mind.

The species held in the Botanic Garden represent an international reference collection. All plants in the garden are labelled with species name, family name, geographical distribution, collector or source, and accession number. Thus the label acts as a key to the information about the species and the labelling of the collections must therefore be maintained to a high standard of accuracy. This requires constant evaluation of the collection by plant taxonomists and updating of the records held at Cory Lodge.

Many of the plants in the Botanic Garden grow in grassy meadows. This very modern feature of gardening in Britain has been in vogue for about ten years but the practice is now nearing its century here in Cambridge. The diversity of species within the flower-rich meadows, such as *Primula veris* (cowslip), is remarkable and adds delight to the garden in early summer. Judicious cutting maintains this diversity of plant life and allows maturation of a fine butterfly fauna.

Reginald Cory gave a major bequest to the Botanic Garden which enabled the eastern section of the garden to be developed from 1951. The structure of the new area of the garden has now been defined by the trees planted within it and the visual impact will develop to maturity in a further fifty years. This section of the garden will then be one century old – venerable and mature with a woodland ride snaking through the middle of the area. The creation of the new section of the garden in the twentieth century has allowed new developments in Botanic Garden style. For example, the garden of scented plants for the blind and partially sighted was one of the first to be designed. Most spectacular of all is a winter garden designed to be more beautiful in the middle of the winter than any part of the garden in the middle of the summer. With its complex and intricate shapes, its glorious colours, and sumptuous textures, it is artistry, transcending gardening. The combination and juxtaposition of stems, foliage and some flowers is entrancing and brightens the gloomiest of winter weather.

The native plants have not been neglected in the Botanic Garden. The ecological mound was erected by John Gilmour in 1965. It consists of a hillock composed of different calcareous substrates – carboniferous and oolitic limestones, chalk and chalky boulder clay – each colonised by appropriate native species. It presents a magnificent spectacle from April to October, showing the beauties of our native flora. In 1997, a perennial border consisting solely of British wild flowers has been constructed to demonstrate the garden-worthiness of these flowers as well as

their value as food resources for the native fauna of bees, butter-
flies and flies. The city of Cambridge developed on the southern
edge of the Fens, an enormous species-rich wetland area. This
habitat is now reduced by drainage for agriculture to a few minute
and scattered areas. In the Botanic Garden a fen has been constructed which demonstrates the complexity of fen vegetation and
its relationship to water depth. These features illustrate the rising
consciousness within the Botanic Garden of its role in the interpretation of the natural world to students and visitors. The
Botanic Garden is an excellent learning environment which can
present complex scientific concepts through the medium of the
plants.

A recent development adjacent to the fen has been a maze
particularly for children. This maze consists entirely of grass –
turf paths are separated by waving golden walls of a New Zealand
species *Oryzopsis leesoniana* giving a beautiful, inviting and non-
threatening structure. It has proved popular with children of all
ages and has seemed remarkably satisfying for undergraduates
during their final examinations.

Throughout the year the Botanic Garden is a visual delight. In
the depths of the winter the skeletons of the trees are exposed in
their magnificence. Particularly striking are the specimens of
Metasequoia glyptostroboides (dawn redwood), a deciduous conifer
with a tall stately presence. This species was known only as a
fossil until its discovery in a Chinese valley in 1941. Our trees
were grown from the first collection of seeds made by botanists of
the Arnold Arboretum, Harvard University, in 1947 and distributed around the world to major botanic gardens. In their fiftieth
year, these wonderful trees are amongst the tallest in the University Botanic Garden and are already rivalling the giant redwoods
(*Sequoiadendron giganteum* on the main ride) which are now nearly
150 years old. The evergreen conifers such as *Sequoiadendron*,
however, give the persistent framework of the Botanic Garden
standing proudly against the winter skies.

Spring advances with the eruption of flowering bulbs underplanting huge areas of the Botanic Garden. The native wild daffodil *Narcissus pseudo-narcissus* gives a natural look to the grassland across the garden and contrasts with the vivid blue of numerous *Scilla* species. An arresting and startling sight in March is *Magnolia sprengeri* var. *diva* with enormous pink globes floating against the bare branches of the *Magnolia* collection.

Early summer bursts into colour across the garden with such delights as the free-standing bed of *Wistaria sinensis* dripping with long blue racemes of sweetly scented pea flowers. The rock garden comes into its own, the smallest specimens covered in brilliant blooms contrasting with the white-grey of the carboniferous limestone. By August the Indian bean tree, *Catalpa bignonioides*, from North America, scents the air of the rock garden with the delicious fragrance of its large foxglove-like flowers, and we move on into autumn with the glories of foliage colours and fruits. All year round, the garden offers colours, textures and superb compositions set amidst the glories of mature trees.

All this has not been achieved lightly. The Botanic Garden is a monument to the brilliance of generations of horticulturists and still requires huge, devoted input from the garden staff with a combination of back-breaking effort and delicate finesse to maintain the standard. Cambridge University Botanic Garden, as you would expect, is an important teaching garden. Generations of horticulturists have been raised here in Cambridge, who now look after the world's great gardens. Today's students, who spend one or two years in the garden, will within the next twenty years be in charge of major collections of plants in Britain and overseas. They receive their training by practical work amongst the 10,000 species under the guidance of and alongside the horticultural staff. There are massive tasks to undertake; for example huge hedges need attention and the Botanic Garden has many. Maintaining a hedge four metres in height and one hundred metres long is not an easy task.

The trainee gardeners are encouraged to pursue their own interests and to develop their individual skills. Thus a lecture on willow and its importance in basket-making inspired the student gardeners to recreate a woven-walled turf seat around a sweet almond. Seats of this type are found in medieval illustrated manuscripts – a woven seat soil-filled, and turfed – what was known in French as a *banque* or 'bank' in English. On such *banques*, light snacks and drinks were consumed, the origin of the banquet.

Botanic Gardens are not simply collections of plants, to be held for their own sake and look beautiful. The function of the Botanic Garden is to provide material for research and teaching within the University of Cambridge and for the rest of the country. In addition to the public glasshouses, for example, are those the general public does not normally see. In these are held research collections. They may be samples of wild populations of plants for genetic studies; for example *Rumex acetosa* collections are currently being studied to explore the relationship between structural variation in the genetic material (the chromosomes) and variation in the environment from which the plants were obtained. Other studies involve such diverse disciplines as physiological adaptation, biochemistry of carbohydrate metabolism, photobiology, and the ecology of woodlands. The Botanic Garden, then, is deeply enmeshed in the research life of the University, and such research may not be of a type which visitors would normally associate with such a beautiful place.

One of the common British birds is the dunnock, *Prunella modularis*, a small insignificant little bird which hops inconspicuously under bushes and hedges. The behaviour of this bird has been the subject of an intensive study within the Botanic Garden by Professor Nick Davis of the Department of Zoology. He has studied the breeding behaviour of this mundane and drab bird and established its startling sex life by observations made in the protected environment of the garden. One sexual variant involves two males and one female. The dominant male is shadowed by a subordinate. This subordinate rushes in to copulate with the female whenever the attentions of the dominant male lapse. Before copulation, the female exudes a drop of sperm, the last ejaculate of the dominant male. The subordinate male may thus father the brood, and if this is so the growing chicks are attended by three parents, an obvious advantage.

This study was carried out within the Botanic Garden where hours of observation in the undisturbed peace of the early morning and late evening were necessary to piece together the remarkable life history of this bird. The Botanic Garden thus provides a protected environment, a resource of enormous value.

Plant/animal interactions are an important area of investigation in modern biology. For example, the preferences of hive bees and bumble bees for flowers are poorly understood. Bumble bees are declining in nature, but without these major pollinators seed production of many agricultural and horticultural crops will decline. Experiments are in progress in the Botanic Garden on pollinator preference which entail students lying with their faces pressed into specially planted flower beds gazing at the foraging activities of bees. A quiet undisturbed environment is necessary for this study also.

The University Botanic Garden has been significant in the history of biological thought. William Bateson, a Fellow of Trinity College, was amongst the first after the rediscovery of Mendel's work in 1900 to carry out confirmatory genetic analyses to confirm and expand our understanding of heredity. This work was done in the Botanic Garden. One plant he chose for his research was the sweet pea (*Lathyrus odoratus*) and the characters he chose to investigate were the colours of the petals. As a result of his studies and those of his collaborators novel genetic phenomena such as gene interaction (epistasis) were established. The Botanic Garden was for sixteen years a major world centre of the developing science of genetics. A living display of Bateson's formative work and ideas is currently being assembled in the public part of the Botanic Garden.

During the period 1922 to 1947 another geneticist, Charles Chamberlain Hurst, worked in the Garden on the genus *Rosa*. Hurst was able from genetic studies to establish the history and origins of the garden roses (the Floribundas and Hybrid Teas) from the ancestral wild species. In the course of his work he made many hybrids. One of these, with flowers of a clear yellow, has been named *Rosa cantabrigiensis*, the rose of the Botanic Garden, and a most delightful spring-flowering plant.

Nine different groups of plants are held in the Botanic Garden as national collections. These are held for reference and for study and involve both woody genera (*Ruscus, Ribes*) and herbaceous genera (*Geranium, Fritillaria*). Such collections can be used in the compilation of biodiversity records anywhere in the world.

One of the most remarkable collections in the Botanic Garden is that of juniper (*Juniperus communis*). Many people have consumed the essence of this species in liquid since it is the flavouring that gives the specific taste to gin. Juniper is historically one of the world's most influential plants. For example, the British domination of its empire in India was due in part to juniper in its form as gin. Gin and tonic was a staple, containing quinine from the South American shrub *Cinchona*, which warded off malaria. It therefore took administrators ten years to die instead of five years.

The Botanic Garden collection of British juniper is amazingly diverse – prostrate forms, arching shrubs and tall trees. So variable are the plants that each individual is recognisable. This diversity is clearly related to environmental conditions. The prostrate plants grow on exposed cliffs in Cornwall and the north-west of Scotland, environments with atrocious weather conditions where gales occur on perhaps two hundred days a year. By contrast the tall trees come from the Cambridge area, in the east of the county where the plants experience warm and dry conditions on chalk grassland. The interaction between the environment and the genetic structure within this remarkable species can be displayed by growing the collections in the common environment of the Botanic Garden.

Botanic gardens worldwide are now linked together into networks. One of the sad linkages, or perhaps now more hopeful linkages, that Cambridge has developed is with Sarajevo Botanic

Garden in Bosnia. Sarajevo Botanic Garden is situated in the centre of the city and was fought over for years during the civil war. The garden was amongst the most important in the Balkans and was devastated by the fighting. There are horrific pictures of mortar bombs destroying this remarkable collection of native and endemic species. Dr Max Walters, a previous Director of Cambridge Botanic Garden, has been co-ordinating the efforts of the botanic gardens of Europe in assembling new collections to repopulate the Sarajevo garden now that a form of peace has returned. International collaboration of plant collection holders in Botanic Gardens is strong and effective.

We may ask what is the need for concern about plants? The reasons lie with biodiversity and the stewardship of the earth for a sustainable future. Most governments are signatories to the Convention on Biological Diversity, the Rio Declaration signed in 1992. Under the terms of the Convention each nation is required to make an inventory of its own biological diversity – all the species that live within its territory – and these species must be protected. Botanic Gardens are a major resource used in the compilation of species diversity. Furthermore, of all living organisms, 80 per cent of them by weight and hence by bulk, are green plants; all other organisms comprise only 20 per cent. Green plants act as the channel through which energy flows to the organisms of the world. Without the photosynthetic activities of green plants energy transfer would cease and death would ensue. Thus we must think about the plants themselves, their maintenance and their continuation through to the twenty-first century, the stewardship of the world's resources.

Ideas of stewardship and of biodiversity and many serious issues in science can be readily explored in a garden such as that of Cambridge. Huge numbers of visitors come each year since it is a beautiful and peaceful place but, while they are in the garden, we hope to teach them, albeit gently, so that they take from the garden messages of many different types. People may come simply to enjoy the beauty, which we encourage. Others come to relax during the testing times leading up to exams: revision in the Botanic Garden can ease stress and tension, more easily than psychological counselling. There are so many elements to the Botanic Garden which are appropriate to children, the disabled, and the whole spectrum of society. There is, inevitably, a refreshment area, within a building whose use is multitudinous – as a lecture room, exhibition centre and workshop. The Gilmour Building is named for the second Director, John Gilmour. Knowledge and food are dispensed here.

There are various methods of teaching and learning. One of the most popular activities in a beautiful garden is art. Throughout the year you will find people drawing and painting in the Botanic Garden. It is very important for us that messages about the natural world are inculcated at a very young age: messages about ecology, about sustainability, about diversity, messages which can also be fun. *Gunnera mannicata* from South America is a major child attraction. The lure of its enormous, sheltering leaves is irresistible and under its canopy children can begin the process of appreciation of the plants of the world. The glasshouses offer other possibilities. On wet days and on dull days in the winter, we can teach about tropical crops and the dependence of humans on plants. Children can see bananas, cocoa pods and also the tree which supplies the raw material for chewing gum. The plants provide an entry to the study of human society, to culture and to art, as well as economics and the sciences.

Plants, like animals, differ enormously in their body plans and the diversity of forms of flowering plants is amazing. We are beginning to unravel the genetic control of how shape is determined. In the Botanic Garden we are devising plantings which illustrate the complex ideas of plant development. The leaf of the garden pea is a powerful visual system in which mutants illustrate gene action in the determination of leaf shape. For example genetic variants affect the presence or absence of tendrils. One mutant form, termed leafless, has all its leaflets converted to tendrils. This mutant is the mainstay of the frozen-pea industry.

Leafless peas are self-supporting in the fields and thus keep the pods away from the ground and from the predatory slugs. An added benefit is a reduced quantity of waste material after harvest. The mode of action of this gene is currently being investigated in cultivated peas but we can see the same effects in action in the evolution of wild species of legumes. Thus we are beginning to understand something about the control of the morphology and development of plants. These complex concepts can be illustrated and illuminated by specific plantings in the Botanic Garden.

The living world involves complex interactions between plants and animals. Plants and animals have lived together for hundreds of millions of years, interacting dynamically with each other. Thus the animal kingdom, the fungal kingdom and the plant kingdom are co-evolving in mutualism, in symbiosis and in antagonism. The Botanic Garden reflects this and projects these ideas to the public. The Botanic Garden is a teaching and research resource, which seeks to put over significant and urgent information through its beauty. The guidelines of resource and amenity which were established by Professor Henslow in the early nineteenth century here in Cambridge are still timely and will increasingly be so as we move to the new challenges of the coming century.

CHAPTER 10

Geophysics in Cambridge: extinct and active volcanoes

HERBERT E. HUPPERT

THE BEGINNINGS

Geophysics is the study of the structure and dynamical evolution of the Earth using the concepts of physics in a quantitative manner. Like many branches of physics, the study of geophysics can be considered to have commenced with Sir Isaac Newton, who occupied the Lucasian Professorship in Cambridge from 1669 to 1701 (the current occupant being Stephen Hawking). Before the time of Newton, our knowledge of the structure of the Earth was rather vague. Aside from knowledge of the transparent atmosphere and neighbouring oceans, extrusion of hot, smelly liquid from volcanoes on the Earth's surface gave rise to the notion of a predominantly solid Earth in which there were interconnected vast subterranean caverns of hot sulphurous material – somewhat consistent with the then current views of Hell.

The quantification of Newton's inverse square law of gravitation allowed the Earth to be 'weighed'. This was done in 1775 by measuring the deflection of a 'vertical' plumb bob by the mountain Schiehallion in Scotland at various distances from it. These measurements allowed the total mass of the attracting Earth to be estimated at 5×10^{24} kg. With the known mean radius of the Earth of around 6,400 km the result suggested a mean density of 4,500kg

m^{-3} (modern measurements give 5,520 kg m^{-3}) in sharp contrast to the measured mean density of 2,500 kg m^{-3} of almost all near surface rocks, including those that make up Schiehallion. The inescapable conclusion is that density within the Earth must vary, with at least some portions over twice the density of that at the surface. The only abundant element of high density is iron; and thus the prevalence of an iron-rich interior is suggested.

Earthquakes, which are often, but not always, associated with volcanoes, are viewed as destructive and to be feared by many; but they are also viewed by geophysicists as an important way of understanding the Earth. In 1761 a long paper appeared in the world's oldest scientific journal, the *Philosophical Transactions of the Royal Society* (of which I am on the current Editorial Board) by John Michel, then Woodwardian Professor of Geology at Cambridge, (Professorial) Fellow of Queens' College and rector of St Botolph's church (all of which he gave up three years later, preferring to get married. Only the Master of a College was allowed to be married until there was a change of statutes in the late 1800s). Michel's suggestion for seismic disturbances was that they resulted from the vaporisation of water when it came into contact with volcanic fire. His opinion was that the formation of small quantities of vapour would lead to a small vibrating cavity, while large quantities would lead to the vapour bursting out of the cavity and travelling through the space (or interface) between different layers.

Modern seismology began at the end of the last century, at about the same time as the phenomenon of X-rays which can see through human bodies were developed in medicine. When a disturbance occurs within or on the Earth, due either to natural or man-made processes, waves, known as seismic waves, propagate away from the source, in a similar way to the acoustic waves which propagate from a speaker's mouth to everywhere in a room to be picked up by each one of the audience. An important difference is that in an elastic medium there are two sorts of waves possible, which travel at different wave speeds, and are called

P (primary) and S (secondary) waves. P waves are quite similar to the compressional acoustic waves in air, while S waves cannot propagate in a fluid. On 18 April 1889 the effects of an earthquake in Tokyo, Japan, were observed one hour and three minutes later in Potsdam and Wilhelmshaven, Germany, half way across the world, as a result of waves which had travelled on the surface of the Earth. We now know that there are a myriad of waves, made up of the P and S waves, which travel a more direct route, through the interior of the Earth, and would have arrived earlier, but were apparently not then detected.

At the turn of the century, a British geologist, Richard Oldham, looked carefully at the travel times of earthquake waves travelling through the interior of the Earth as a function of distance (along the Earth's surface) from the source, which could be expressed as an angle (between 0 and 180°). He believed he could account for the observations if the Earth had a homogeneous (heavy) core whose radius was about 0.4 times the Earth's radius, that is, about 2,550 km. In this core, he suggested, the seismic wave velocity was considerably less (by almost a factor of two) than in the surrounding material. He argued that seismic waves which entered the core at an oblique angle change their direction of propagation, in the same way that a light beam is refracted as it passes into a medium, such as glass or water, in which its speed is less than in air. A spherical core – why should it be anything but spherical? – would bend the seismic waves entering at different angles, and they would be bent again on leaving. He suggested further, by considering S waves, which do not propagate in a fluid, that the core was fluid. What the Earth is made of, or for that matter the determination of the constituents of any part of it, requires more, and different, information. The argument that at least part of the core was fluid was put forward by the remarkably talented Cambridge geophysicist Sir Harold Jeffreys, of whom more will be said below. Jeffreys based his arguments on the rigidity of the core, which could be determined both from seismic wave velocities and from tidal motions of the solid Earth.

The final, and courageous step was taken by Inge Lehmann, sadly the only female geophysicist to enter our story. Born in 1888, Lehmann died at the age of 104, the holder of a coveted Foreign Membership of the Royal Society (of London) and admired by all who knew her. With careful theoretical calculations, guided by detailed seismological evidence, she wrote in 1936: 'A hypothesis will here be suggested which seems to hold some probability, although it cannot be *proved* [my italics] from the data at hand . . . that inside the core there is an inner core in which the velocity is larger than in the outer one', and that inner core is solid. Not all geophysicists agreed with Lehmann at the time and it was said that her results were based on 'exacting scrutiny of seismic records by a master of a black art'. Her hypothesis was not generally accepted until the 1960s when extra, and supportive, evidence came from analyses of the Earth's free oscillations generated as a result of the massive earthquakes in Chile in May 1960 and in Alaska in March 1964.

Our present view of the structure of the Earth is a solid inner core of 1,221 km surrounded by a vigorously convecting, liquid outer core to a radius of 3,486 km and a predominantly solid, silicate mantle to a mean radius of 6,371 km.

A SCHOOL IN CAMBRIDGE

However, I am allowing the scientific description to outrun the chronology and especially Cambridge chronology. Let me go back to 1891 when Hugh Newall, the Professor of Astronomy, had built the fine Victorian House, Madingley Rise, off the Madingley Road, just over half a mile north-west of the centre of Cambridge. In 1898 there was a total eclipse of the sun visible from northern India, and Newell, with his wife, journeyed to one of the two officially designated observation sites at Pulagaon. He met there, and became good friends with, Gerald Lenox-Conyngham, who had been to the Royal Military Academy at Woolwich and who was then Assistant Surveyor-General of the

Indian sub-continent (and whose wife was in charge of the domestic arrangements at Pulagaon). Lenox-Conyngham was engaged primarily with astronomical measurements and determining the local value of gravity by measuring the period of an oscillating pendulum. A knowledge of the local value of gravity and using Newton's law of gravity allowed Lenox-Conyngham (and others) to determine the local density of the rocks beneath the surface. For this work Lenox-Conyngham was elected a Fellow of the Royal Society in 1918 and knighted in 1919. He retired from the Indian Survey in 1920 and left India in May 1920 at the age of fifty-five, considering a peaceful retirement in Oxford.

For the first twenty years of this century there had been various suggestions that practical geodesy needed strengthening in Great Britain and that a Geodetic Institute should be set up. The scientific necessity was recognised, but, as often, there were financial difficulties. In 1920 the Vice-Chancellor of the University of Cambridge proposed that, if finances could be secured, a Geodetic and Geophysical Institution be established within the University, and a small committee which included Professor Newell was set up to investigate the possibility. Newell was also a Fellow of Trinity, the Bursar of which wrote to the Vice-Chancellor early in 1921 saying that: 'if the University made satisfactory arrangements for . . . research in geodesy, the College would be prepared to assist [with the finances]'. Newell then turned to his friend Lenox-Conyngham, who was elected to a Fellowship of Trinity and a Readership of the University on setting up the School of Geodesy on 7 October 1921. Thirty-four years later, the Department of Geodesy and Geophysics, as it was then to be known, most appropriately moved into Newell's old home, Madingley Rise, which still houses some of the most active volcanoes of British geophysics.

Lenox-Conyngham continued his work on determining the local value of gravity by using swinging pendulums, taught undergraduates and expanded his School by hiring (amongst others) two of the great British geophysicists: Harold Jeffreys and Teddy Bullard. In later life, Lenox-Conyngham was once asked how was it that without having attended a university (as an undergraduate) he could become: a Fellow of the Royal Society; a Fellow of Trinity; and a Professor (Reader actually) in Geophysics. 'Easy . . . I started the subject' was his reply.[1]

SIR HAROLD JEFFREYS AND THE EARTH

Harold Jeffreys was a lecturer in mathematics at Cambridge of considerable standing, when in 1931 he was elected to a Readership in Geophysics and became part of Lenox-Conyngham's 'team'. He was a highly skilled mathematician who applied the concepts and methods of classical mechanics to unlock the secrets of the interior of the Earth. He once said: 'If it takes complicated mathematics to understand the Earth, blame the Earth; and not me.' After showing that the core was fluid, as described above, Jeffreys turned his attention to the propagation of seismic waves through the interior of the Earth. If the velocity of waves (either P or S) is known as a function of radius within the Earth, then the travel time of the various rays between an earthquake and an observation point can be calculated. There can be many such rays. With such knowledge any earthquake can be located in both space and time by recording arrival times of waves at a number of stations on the surface of the Earth. Jeffreys, in collaboration with his young research student from New Zealand, Keith Bullen, who was to become Head of the Department at the University of Sydney from which I obtained my undergraduate degree, analysed enormous amounts of data from a large number of earthquakes collected on a routine basis to determine travel timetables, which came to be known as the Jeffreys–Bullen or J–B tables. The first J–B tables were published in 1935 and were then upgraded in 1940. They continue to be used up to present time, with very little modification, to locate earthquake epicentres world-wide. Dealing with the large amount of data, and the inherent errors, led Jeffreys to initiate important aspects of probability theory; but this is another story.

Jeffreys's main influence, on countless geophysicists, was his celebrated book *The Earth*, the first edition of which appeared in 1924 and the seventh and final edition in 1976. In it he described the interior structure and history of the Earth as he saw it. The third edition was reviewed as 'probably the most frequently quoted book on the interior of the Earth'. Already in the introduction to the first edition Jeffreys stated one of the most important and challenging aspects of geophysics: 'the problem . . . is to make physical [and chemical, actually] inferences over a range of depths of over 6,000 km from data determined only for a range of 2 km at the outside'. Somewhat surprisingly, given their importance, *The Earth* says very little about volcanoes, although Jeffreys recognised (in an ungrammatical sentence) that 'the formation of volcanoes . . . in a liquid or partially liquid state, to [*sic*] the surface or near it requires explanation'. The discussion which follows has now been largely superseded; and more modern ideas on this matter will be presented near the end of the chapter.

Jeffreys also made important contributions to our understanding of the dynamics of both the atmosphere and the oceans, central areas of geophysics, descriptions of which are lamentably absent from this presentation. But the Earth is a big place and one can't describe all of its workings in one chapter. The absence of any discussion of meteorology and oceanography also precludes any mention of the highly creative theoretical and experimental research of one of the central founders of fluid mechanics, Sir Geoffrey Taylor, who spent the seventy years between 1905 and 1975 working almost exclusively in Cambridge.

SIR EDWARD BULLARD AND CONTINENTAL DRIFT

Sir Edward Bullard, or Teddy as he was known by almost all, was one of Lenox-Conyngham's early recruitments. It has been written that Teddy 'will be remembered as one of the major figures in the development of the Earth Sciences during the twentieth century' and definitely one of the most colourful characters. A one-hour lecture could easily be given consisting just of anecdotes about his doings, sayings and buccaneering style. One, motivated by a famous photograph, will suffice here. I was a graduate student in the mid-1960s in the Institute of Geophysics and Planetary Physics (IGPP) which was housed in a beautiful, purpose-built, redwood building overlooking the cliffs of La Jolla, California. One of the many imaginative aspects of the building, which was partially designed by Judy, the wife of the then Director Walter Munk, who himself is a very creative geophysicist and a colourful character in his own style, was that offices had no numbers attached to them. To differentiate office from office, a large photograph of a distinguished geophysicist, generally in informal dress, was hung on each door. My office was adorned by a life-size photograph of Teddy taken while he was measuring gravity in the East African Rift Valley. Teddy was a much welcomed visitor to IGPP. I often heard him approach with visitors, some quite distinguished, and he would always say something like: 'Aren't I well dressed – I choose the hat to match perfectly the rest of my clothing!' (He was dressed *only* in a hat.)

Lenox-Conyngham was at first uneasy about hiring this tall, larger-than-life character who, much later, was to say: 'I always had the feeling that I was really the Wizard of Oz, the best wizard there was around. . ., but behind it all a bit of a fraud.' At some time during Bullard's early days, Lenox-Conyngham was sharing his doubts about Teddy with Lord Rutherford over port at Trinity, where they were both Fellows. Rutherford's view was: 'I'll tell you what, Conyngham, he's a damn sight cleverer than you are.'

Teddy was trained as an experimental physicist working in the Cavendish under the direction of two Nobel Laureates and Peers of the Realm: Lords Blackett and Rutherford. He used his extensive experimental abilities to investigate (and often initiate) a number of fundamental areas of geophysics. Very early in his scientific career he realised that although approximately 75 per cent of the Earth's surface was covered by oceans, virtually nothing was known about marine geology, in contrast to continental geology, although admittedly the latter was much easier (and far less expensive) to study. After the Second World War, due to extensive work at sea by Teddy and companions in Cambridge and other groups in America, it gradually became clear that the geology of the ocean floor differed considerably from that of the continents. At sea, all the hard rocks are basalts, indicating a volcanic origin; all the hills are volcanoes, in contrast to, for example, the massive, non-volcanic Himalayan range consisting almost entirely of granite; the covering sediments are considerably thinner than those found on the continents or the surrounding continental shelves; and, most important, all rocks, no matter where or how the ocean floor was dredged, were younger than 100 million years, in contrast to continental rocks which are up to 4,000 million years old. The inescapable conclusion was that the ocean floor was much younger than, and formed in a different way from, the continents. Further, gravity measurements indicated that rocks well below the ocean floor were not just sunken continents covered by lavas from volcanic eruptions in the ocean (or elsewhere) and recent sediments. In addition there is the curious fact, that there are relative highs, or long, quite thin ridges, positioned rather accurately half-way between the two continental edges.

Almost every schoolchild who has sat (bored) through geography lessons has wondered about the apparent fit between the west coast of Africa and the eastern coast of South America. The 'fit' has been commented upon since at least the time of the great English essayist Francis Bacon in 1620 and advocated strongly (using some incorrect arguments) by Alfred Wegener in 1912. Teddy, stimulated in part by contemporary themes we shall discuss in a moment, decided to investigate the fit quantitatively. He introduced two essential constraints. First, he matched the continents *at the edge of the continental margins*, rather than at the shoreline, explicitly taking into account a difference in position which varies quite considerably along the coastline. He did this because geological investigation had indicated that the continental shelves, often covered by relatively thick sediment, where oil and gas has so successfully been found, are part of the adjoining continent, rather than part of the oceanic floor. Second, he applied a powerful result of spherical geometry, known as Euler's Theorem and occasionally used by mathematicians since 1776, that states that the *rigid* motion of a (partial) covering of a sphere to another location is equivalent to a rotation of a particular angle about a specific pole of rotation. This study, with his Madingley Rise colleagues, Everett and Smith, led to the wonderful fit which is reproduced in many textbooks on the Earth. This result, and Teddy's dramatic and showmanship manner of presenting it, had considerable influence and suggested that the relatively older continents once made up one much larger land mass before the continents drifted apart to form the relatively younger oceans between. This was the quantification of an important area of geophysics called *continental drift*.

At about the same time, Fred Vine and Drum Matthews, a PhD student and his supervisor at Madingley Rise, presented a new and powerful interpretation of measurements of the magnetic field in rocks on the ocean floor. Such measurements had begun to be collected near the end of the Second World War by Blackett, Teddy's own PhD supervisor. Vine and Matthews employed the well-known result that all magnetic substances have a particular temperature known as the 'Curie point'. On being heated above their Curie point, substances take on any prevailing magnetic field; as they cool below the Curie point the direction of the magnetic field is frozen into the substance forever (unless it is either once again raised above its Curie point or significantly altered, such as by repeatedly subjecting it to severe blows). Vine and Matthews, analysing magnetic surveys they had undertaken in the Indian Ocean (following extensive, earlier surveys by others in the sea floor off California) showed that the map of frozen-in magnetic fields consisted of parallel lineations bearing no relationship to the topography of the local sea floor but aligned with the axes of the giant ridges. Vine and Matthews proposed that new, hot lava was being continuously (on a geological time scale) extruded from undersea eruptions along the length of the mid-oceanic ridges and then spread sideways at between 1 and 10 cm/yr – about as fast as one's fingernails grow. With each reversal of the magnetic field of the Earth a separate 'track' of magnetisation was laid down. This turned attention from *continental drift* to *seafloor spreading*.

It is (at least to me) an interesting, but a not unusual twist of science that when these suggestions became known to a quite senior marine geophysicist in California he said: 'What rubbish. I am the world's expert, and holder of the greatest amount of data, on magnetic field observations on the ocean floor. If the Vine–Matthews hypothesis had any validity it should be clear in my own data.' He then looked; and it was obvious.

DAN MCKENZIE: PLATE TECTONICS AND COMPACTION

The fundamental concepts were then finally completed by one of Teddy's very brightest, and definitely most successful, graduate students, Dan McKenzie, who is very much one of today's active volcanoes. Using a sophisticated computer programme devised by his friend, Bob Parker, who had also just graduated with a PhD under Teddy's direction, McKenzie and Parker showed from the motions of earthquakes that large portions of the surface of the Earth do actually move as rigid, spherical caps or plates, with an average depth of order 75 km, with little seismic activity within the plates. So arose the final terminology of *plate tectonics*, whose introduction in the mid 1960s represented the largest revolution ever in the concepts and consequences involved in the Earth Sciences. Somewhat surprisingly, the transition was very rapid by scientific standards – especially given the time scale of geology. It has been said that in the early 1960s no one who even suggested that there may be something in the idea of continental drift could hope to be hired in a reputable North American university, but by the late 1960s no such position was available to anyone who didn't believe totally in its concepts.

There were, of course, many further details of plate tectonics to be worked out; and this was done over the next thirty-five years, in part by McKenzie himself. We are now in the amusing situation where, in contrast to when Teddy started thinking about marine geophysics, we have a better understanding of oceanic plate motions than continental geology.

What drives the plates? The simple and superficial answer is convection in the mantle below, driven by two processes. First, the loss of heat initially contained in the Earth. Very roughly, there are indications that the temperature of the interior of the Earth has decreased 400° C in its 4,500-million-year history. Second, heat is generated by the decay of radioactive elements within the Earth. The convective motions are loosely akin to those when porridge with a thin crust is heated from below. But there are still many unanswered questions, including the following. To what depth does the convection (in the solid rock) extend? Some say from just below the crust down to 670 km; others say down to the base of the core, 2,885 km below the Earth's surface. How temporarily and spatially varied is the motion? Why are the plates the size and shape we observe? All these questions, and other related ones, are being actively researched at the moment, frequently involving the most powerful computers in the world. Even though we don't fully understand all about the driving mechanisms of plate tectonics, the concepts have a firm foundation. As the powerful mathematician of the last century, Oliver Heaviside, said in reply to critics of his new calculus which produced the correct answers, even though he, and more importantly, his critics did not understand why: 'Should I desist from eating, just because no-one understands the workings of my digestive tract.'

McKenzie has gone from strength to strength; and has won

major international awards for his scientific contributions almost every year in the last fifteen. One of his contributions came from asking, as Jeffreys had previously, how does liquid, formed in small gaps in the solid Earth, escape to produce the enormous amount of volcanism at the Earth's surface? In a study which has been highlighted on a special Horizon TV programme, McKenzie showed that liquid in the space between the exceedingly small crystals or grains that make up most of the rocks of the Earth is interconnected and can flow along the boundaries between the grains. The liquid forms because as the solid rock slowly moves it comes to areas where the local temperature (at that local pressure) exceeds the melting temperature of some of the constituents of the rock and these components therefore begin to melt. The melt can then be squeezed out of the solid matrix by the overlying pressure by a process known as *compaction*. The small trickles of magma or molten rock, at around $1,000°$ C, can be separated from the solid matrix and wind their way upwards in channels of increasing width, until, by processes that are not yet fully understood, they erupt at the surface in volumes as large as 10 km^3 and instantaneous rates of up to 1 km^3 / day – fast enough to fill a fair-sized lecture hall in about a second.

GEOLOGICAL FLUID MECHANICS

One morning, in September 1979, the telephone rang in my office in the Department of Applied Mathematics and Theoretical Physics and the voice at the other end introduced himself as Steve Sparks, a volcanologist in the Department of Earth Sciences, saying: 'Dan McKenzie told me that you knew some fluid mechanics; is that true?' We arranged to meet; we talked about his research in volcanology and mine in fluid mechanics (mainly applied to oceanography and meteorology); and roughly, neither of us understood a word the other was saying. But I quickly realised that Steve was an outstanding scientist and a very nice individual, who was working in a potentially exciting area; and he realised the power that a quantitative understanding of fluid mechanics could bring to his research. Put briefly we (and our families) became good friends and together we initiated the new field of *geological fluid mechanics*, some of whose results I will quickly discuss here.

First, however, I would like to digress for a moment, motivated by my statement that Steve is a very nice individual. The Australian, Nobel-laureate developer of penicillin, Lord Florey, once said: 'I'd work with the devil, if he had something to teach me.' Florey's Nobel-prize-winning student, Sir Peter Medawar, once wrote, I am sure with his teacher's comment in mind: 'I don't like to work in the lab. with someone whom I am not equally pleased to socialise with on Sundays.' Here is not the place to debate these two extreme but interesting views. Let it suffice to say I still enjoy seeing Steve and his family on Sundays. But back to science and in particular the work on the volcanic aspects of geological fluid mechanics which have been developed in Cambridge over the last twenty years or so.[2] Readers interested in a more detailed discussion might consult the references listed in notes 2–7 below.

Magma chambers

As discussed previously, liquid rock, or magma, rises to the Earth's surface by the process of compaction. It is believed that it may do this by trickling through a series of thin contorted channels or by migrating in batches as much as 1 km^3 in volume. Just a few kilometres beneath the surface the melt often accumulates in storage reservoirs, known as magma chambers, which exist beneath all volcanoes, both on land and (the much more numerous ones) at the bottom of the ocean. These magma chambers, full of hot turbulently moving melt with suspended crystals, act as the 'powerhouse' for all volcanic processes. The chambers range in size from a fraction of a cubic kilometre to a few thousand cubic kilometres (a few tens of kilometres in horizontal extent and a few kilometres in the vertical). Magma may repose in one or more such chambers during its ascent towards the Earth's surface. The repose time for a particular fluid element can range from as little as a few hours through to effectively forever, the latter limit occurring if the magma cools sufficiently while in the chamber to solidify there. In general, however, the repose time stretches from years to thousands of years or much more, during which time the magma cools, partially solidifies and evolves chemically and physically. Fluid-mechanical thinking is making a large contribution to understanding the various processes that occur in magma chambers. In addition, many of the new processes considered, motivated by a desire to understand the natural workings of volcanoes, have direct industrial application. Thus, for example, the solidifying of a multi-component magma and the resulting motion of the remaining fluid, in the form of convection, has parallels with the industrial processing used in the manufacture of steel ingots.

Dry fissure eruptions

As is well known, volcanic eruptions display a range of fascinating and at times awe-inspiring features. This subsection will concentrate on the eruption of magmas of relatively low viscosity (or stickiness) and low volatile content, which is the case when the magma comes from great depths within the Earth, as occurs, for example, beneath Kilauea in Hawaii and Mount Etna in Italy. The magma then tends to be extruded in a non-explosive continuous fashion, often from a long fissure which produces a flow of lava known as a 'curtain of fire'. The subsequent development varies from eruption to eruption. In some cases the height of the curtain of fire and the flow rate decrease and the eruption ceases, generally within a fraction of a day. In other eruptions, after a few hours, a second phase is gradually initiated in which there is a decrease in the length of active fissures accompanied by a concentration of the height of fountaining at certain points along the fissure. If the eruption continues, the flow of lava can become localised to only a few surface vents, around which volcanic cones are gradually built up.

Aside from the initial geometry, the change in temperature of the rising magma and the surrounding solid rock plays an important role throughout the eruption. When magma first fills the fracture, known as a dyke, it initially solidifies against the cold channel walls. Continued solidification may eventually block the channel, which tends to end the eruption at that site even though the driving pressure in the magma chamber may remain substantial. However, the continual supply of heat to the walls of the magma flowing from the chamber may eventually exceed the possible conductive transfer into the surrounding rock. Initial solidification will then be halted and the walls subsequently melted. The width of the fissure then continues to increase until the magma supply diminishes. Which style occurs depends on the initial width and length of the dyke, the driving pressures and the initial temperature difference between the magma and the surrounding rock. All of these effects have been quantitatively examined[3] and the results have increased our understanding of the field data.

Explosive eruptions

In other parts of the world the chemical composition of the magma is such that it is quite viscous (sticky) and it includes a small amount of dissolved water – generally less than a few per cent (by weight), but this is sufficient to have considerable impact and lead to explosive eruptions,[4] as occurred at Krakatoa in 1883 and Pinatubo in 1991. Generally, as the magma cools and crystallises it forms anhydrous crystals (i.e. none of the water is taken up by the crystals). Thus the relative amount of dissolved water remaining in the magma increases, until it reaches saturation. (This happens quite easily; magma is not able to dissolve much water.) Thereafter the water exsolves and forms small bubbles of relatively low density. If this happens while the magma is still in the chamber the pressure can increase dramatically; and trigger an eruption. Alternatively, the magma may have already begun to rise to the surface. Now the continuing decrease in pressure causes the bubbles to exsolve – just like when removing the cork from a champagne bottle. The small, light bubbles rise and, because the magma is so viscous, they take much of the magma with them. As the magma and bubbles rise in unison, the pressure decreases further. The amount of vapour bubbles increases until at a volume fraction of around 75 per cent the mixture behaves like a foam (rather than a fluid). A short distance beyond that the gas content is so large that the behaviour is akin to a high-speed gas flow taking along many small ash particles (all that remains after disruption of the once continuum magma). This flow can become supersonic – move faster than the *local* speed of sound – and then vent into the atmosphere at speeds of between 300 and 700 feet per second as a hot, turbulent, gas-enriched, particle-laden plume which penetrates many tens of kilometres into the atmosphere. The physics of such eruption columns is described in the next subsection.

Volcanic plumes

The greatest height a plume has penetrated the atmosphere this century is 45 km (Bezymianny, in Russia in 1956). While the gas jet at the base of the plume has considerable vertical velocity (and momentum) this is nowhere near sufficient to allow it to rise 45 km. The potential energy is *not* gained from the kinetic energy at the source, but indirectly from the thermal energy in the erupted plume, as follows. The hot, (ash) particle-enriched gas plume erupts turbulently into the base of the atmosphere in the form of a jet, whose large-scale turbulent eddies engulf surrounding air and mix it in with the plume. The small, hot particles readily transfer their heat to the engulfed, relatively cold air which can thus become less dense than the surrounding air (even *after* taking into account the excess contribution to the bulk density made by the relatively (very) heavy particles, whose density is roughly three orders of magnitude larger than that of air). Because of this decreased bulk density, the jet has an increased upward force acting upon it. In this way the plume can penetrate to great heights into the atmosphere by the gradual transfer of the thermal energy in the ash particles.

The currently best quantitative model of such an eruption was developed in the mid-1980s by a then student of mine, Andrew Woods, now Professor of Mathematics in Bristol, whom I set the task of reviewing models of eruption columns and who evaluated for himself a comprehensive sophisticated model.[5] This style of eruption is called a Plinean eruption, as described by Pliny the Younger after the eruption of Vesuvius in AD 79. The plume width increases with height as surrounding air is engulfed, while the velocity of penetration and temperature difference decrease, all in a way which we can now quantitatively evaluate. Because the density of the atmosphere decreases with height, eventually the plume must reach a height where its (bulk) density is no longer less than that of the surrounding air, and so it can go no higher. (Strictly speaking, the plume penetrates slightly higher, driven by the momentum in the plume, but it then falls back to almost this same level due to its relatively larger density at higher levels.) The plume, along with its particles then penetrates sideways into the atmosphere to produce a large 'umbrella cloud' from which the small ash particles slowly settle. These clouds can cause enormous problems for aircraft, as evidenced by the chilling story of the British Airways pilot who controlled a 20,000-ft free descent before being able to restart his engines which had been choked with ash after flying through the volcanic debris of Mount Galungung in south-east Asia in 1982.

The distributed ash in the atmosphere can also greatly influence the weather on a global scale. The whole Southern Hemisphere experienced beautiful daily sunsets for almost a year after the eruption of Pinatubo in 1991. On a larger scale, North America suffered a severe crop failure after the eruption of Tambora, Indonesia, in 1815, leading to a famous book by Dr and Mrs Stommel entitled *Volcano Weather: The Story of 1816, the Year without a Summer*. In a less-well-documented but fascinating account, a religious Cambridge geophysicist (working at Madingley Rise) teamed up with a Cambridge metallurgist to argue that the violent eruption of Santorini in the Mediterranean was exactly at the time of the famous seven years of famine in Egypt.[6] Usually, they argue, in those times the Nile annually overflowed its banks to supply the water needed for crops and cattle. Due to the enormous amount of ash deposited in the atmosphere by the Santorini eruption, the weather would have been completely altered (for seven years) and there could have been an acute shortage of water. On the cause and interpretation of Pharaoh's dreams they are strangely silent!

But back to present-day descriptions. The observant reader will have wondered (correctly) why in the above description of plume motion the bulk density of heavy particles plus hot air must exceed that of the atmosphere at the base of the plume. The answer is that it need not always be so; a different, and important, style of eruption then ensues. Above a critical mass of ash particles, although the exiting jet has an upward momentum, it cannot engulf (and heat) sufficient air to continue propagating upwards. The 'plume' falls back to the ground and travels along it, in what is called by geologists a 'pyroclastic flow'. Both styles of eruption can occur from the same volcano – after some twenty-four hours the Vesuvian eruption produced a pyroclastic flow, which is what killed most of the population of Pompeii and nearby Herculaneum (and Pliny's uncle).

Ash-laden pyroclastic flows can travel enormous distances (in excess of 100 km) at considerable speeds (in excess of 100 metres per second). Given the conditions at the source of the eruption – the mass and momentum of the gas at exit and the particle concentration – it is now possible to calculate the characteristics of the resulting flow: propagation velocity along the ground, final extent of the flow and the variation of the thickness of ash deposited from the turbulently moving flow. This form of calculation is called a forward calculation by earth scientists: given the input (initial) conditions, calculate what happens next. Often, the real problem of interest, however, is an inverse problem: given only (aspects of) the final result, calculate the conditions which gave rise to this result. In this particular case: given the observable deposit from an (ancient) pyroclastic flow, calculate the size and parameters of the eruption. The mathematics to answer this particular situation has only recently been developed[7] and was applied to analyse one of the largest pyroclastic flow eruptions in the last 10,000 years: the eruption of Taupo, New Zealand in AD 186. Our calculations indicate that the total flow rate of gas and solids was around 40 cubic kilometres per second for around fifteen minutes. The near-vent solids concentration was only about 0.3 per cent by volume and led to a pyroclastic flow of about 1 km thick which travelled outward from the vent with a typical speed of 200 metres per second. The low particle concentration was a new (and somewhat controversial) finding, which the science writer for the *Cambridge Evening News* headlined as: 'Cambridge Professor proves ancient eruption nothing but hot air.'

Since pyroclastic flows, world-wide, represent one of the largest natural disasters and have contributed to thousands of millions of dollars' worth of damage to property this century, they warrant further serious quantitative analysis and evaluation.

THE INSTITUTE OF THEORETICAL GEOPHYSICS

In the late 1980s it was thought that theoretical geophysics, particularly seismology, needed strengthening in Great Britain. It was decided by a committee set up by the then University Funding Council to start an Institute of Theoretical Geophysics in the University of Cambridge. The Institute was to be housed in both the Department of Earth Sciences (which by now had taken over, or in, the Department of Geodesy and Geophysics) and the Department of Applied Mathematics and Theoretical Physics. Although quite common in North American universities, such a formal partnership between two separate Cambridge departments had not happened before. There was initially some apprehension shown by the Heads of the two departments, but they both wanted to see the new Institute thrive and worked well together towards this aim.

I was appointed to the Directorship of the Institute in June 1989 and the Institute opened its doors on 1 October 1989, a day which is celebrated annually with a birthday luncheon by its staff. As with the School of Geodesy, Trinity College has helped financially, through its Foundation the Isaac Newton Trust, though not with a Fellowship for the Director – I am a Fellow of King's. The most recent assistance has been with the purchase for £1.2m of an extremely powerful computer for the Institute. It will be devoted entirely to calculations in geophysics. Its speed is such that it can evaluate in considerably less than a second *all* the calculations a scientist, such as Oldham or Jeffreys, could carry out in a lifetime (if they did nothing else). Fortunately, or unfortunately, depending upon your outlook, the thinking power and creative ability of humans is not mirrored in these powerful machines. Combining these two attributes, however – creative imagination and computational power – we are in a wonderful position to learn more about the world around us, to the benefit of us all.

It is a pleasure to acknowledge the helpful comments on an earlier draft of this essay given to me by my colleagues and friends George Batchelor, Ross Kerr, Dan McKenzie, Walter Munk, Howard Stone, Stan Tick and Stewart Turner.

NOTES

1 Peter Bottomley MP, his grand-nephew, private communication.
2 H. E. Huppert, 'The intrusion of fluid mechanics into geology', *Journal of Fluid Mechanics.* 173 (1986), 557–94.
3 P. M. Bruce and H. E. Huppert, 'Thermal control of basaltic fissure eruptions', *Nature* 342 (1989), 665–7 with front cover photograph.
4 A. W. Woods, 'The fluid dynamics and thermodynamics of eruption columns', *Bulletin of Volcanology.* 48 (1986), 245–64.
5 A. W. Woods, 'The dynamics of explosive volcanic eruptions', *Reviews of Geophysics.* 33 (1995), 495–530.
6 R. S. White and C. J. Humphreys, 'Famines and catyclismic volcanism', *Geology Today*, 10 (1994), 181–5.
7 W. B. Dade and H. E. Huppert, 'Emplacement of the Taupo ignimbrite by a dilute turbulent flow', *Nature*, 381 (1996), 509–12.

CHAPTER 11

Cambridge spies: the 'Magnificent Five', 1933–1945

CHRISTOPHER ANDREW

Spies are a Cambridge tradition. Elizabeth I's Secretary of State, Sir Francis Walsingham, a graduate of King's, ran and largely financed a small, underpaid secret service which uncovered the Ridolfi and Babbington plots, provided the evidence which sent Mary Queen of Scots to the scaffold, and finally drove Walsingham himself to the verge of bankruptcy. Among Walsingham's spies was Corpus Christi's most famous poet and playwright, Christopher Marlowe, who absentmindedly left Cambridge without paying his college bill and was killed in a tavern brawl at the age of twenty-eight while on Her Majesty's secret service.

The most celebrated of the Cambridge recruits to the twentieth-century intelligence community have been the codebreakers. When Bletchley Park, the Second World War signals' intelligence centre, decided it needed an influx of 'men of the professor type' to crack the German 'Enigma' machine cipher and solve other cryptanalytic conundrums, it turned chiefly to Cambridge. One-third of the King's Fellowship served at Bletchley during the Second World War, among them Alan Turing, the chief inventor in 1943 of the world's first electronic computer, codenamed 'Colossus'. Some of the Cambridge dons at Bletchley, arguably the most successful intelligence agency in modern history, found themselves upstaged by undergraduates. Sir Harry Hinsley, later

Master of St John's, Vice-Chancellor and official historian of British intelligence during the Second World War, was recruited in the autumn of 1939, just as he was about to begin work for Part II of the Historical Tripos. Alan Stripp, co-editor with Sir Harry of a volume of reminiscences on Bletchley Park, was even younger. He was recruited in the middle of the war soon after winning a Classics scholarship to Trinity.[1]

Cambridge spies have been remarkable for their diversity as well as for their quality. Most universities around the world recruit only for the intelligence services of their own country. Cambridge, however, has a more cosmopolitan tradition. The KGB[2] believed that the ablest group of foreign agents in its history were five young Cambridge graduates recruited in the mid-1930s: Kim Philby, Guy Burgess, Anthony Blunt and John Cairncross of Trinity College, and Donald Maclean of Trinity Hall. After the release of the popular Western *The Magnificent Seven* in 1960, the Centre (KGB headquarters) began to refer to them as the 'Magnificent Five'.[3]

The first of the Five was Kim Philby, who graduated in June 1933 with upper second-class honours in economics and 'the conviction that my life must be devoted to Communism'. It is not difficult to understand some of the reasons for Philby's alienation from the British social and political system of his day: among them the suffocating snobbery of the class system; the complacency of the establishment amid the misery of mass unemployment; the apparent collapse of the parliamentary road to socialism after Ramsay Macdonald's 'betrayal' and the Labour rout in the 1931 election; and the feebleness of the British response to the growing menace of German Nazism and Italian Fascism.[4] What is far more difficult to comprehend is Philby's conviction that the brutal dictatorship of Stalin's Russia was overwhelmingly superior to the flawed parliamentary democracy of the United Kingdom.

What seduced Philby and the rest of the Five, however, was not the reality of the Stalinist regime but the myth-image of the world's first worker-peasant state courageously constructing a new society for the benefit of all. The emotional appeal of this myth-image was so powerful that it proved capable of surviving even first-hand experience of the Soviet Union by many of those whom it seduced. Malcolm Muggeridge, perhaps the most percipient of the British correspondents in Moscow during the mid-1930s, wrote of the idealistic left-wing pilgrims to Stalin's Russia:

> Their delight in all they saw and were told, and the expression they gave to this delight, constitute unquestionably one of the wonders of our age. There were earnest advocates of the humane killing of cattle who looked up at the massive headquarters of the OGPU [as the KGB was then known] with tears of gratitude in their eyes, earnest advocates of proportional representation who eagerly assented when the necessity for the Dictatorship of the Proletariat was explained to them, earnest clergymen who reverently turned the pages of atheistic literature, earnest pacifists who watched delightedly tanks rattle across Red Square and bombing planes darken the sky, earnest town-planning specialists who stood outside overcrowded ramshackle tenements and muttered: 'If only we had something like this in England!' The almost unbelievable credulity of these mostly university-educated tourists astonished even Soviet officials used to handling foreign visitors.[5]

On his last day in Cambridge in June 1933, Philby went to seek the advice of the Communist economics don, Maurice Dobb, on how best to work for Communism. Dobb gave him an introduction to a Communist front organisation in Paris, which directed him to Vienna where Philby spent most of the next year working for the International Workers Relief Organisation and acting as a courier for the underground Austrian Communist Party.[6] While in Vienna he met and married a young Communist divorcee, Litzi Friedmann, with whom he had a brief but passionate love affair which included his first experience of making love in the snow ('actually quite warm once you get used to it', he later recalled).[7] The first to identify Philby's potential as a Soviet agent was Litzi's friend Edith Suschitsky, a Soviet agent unimaginatively code-named EDITH, who moved to London early in 1934, began work as a successful children's photographer and married an English doctor, Alex Tudor Hart.[8]

In May 1934 Kim and Litzi Philby returned to London. A month later Edith Tudor Hart took him to a meeting in Regent's Park with Dr Arnold Deutsch, his first and most remarkable KGB controller, whom Philby knew as OTTO. According to a memoir by Philby in the KGB archives, Deutsch told him, 'We need people who could penetrate the bourgeois institutions. Penetrate them for us!'[9] Philby's first codename, given him immediately after his meeting with Deutsch, had two versions: SÖHNCHEN in German, SYNOK in Russia – both roughly equivalent to SONNY in English.[10]

Half a century later, Philby still remembered his first meeting with OTTO as 'amazing':

> He was a marvellous man. Simply marvellous. I felt that immedi-
> ately. And [the feeling] never left me . . . The first thing you
> noticed about him were his eyes. He looked at you as if nothing
> more important in life than you and talking to you existed at that
> moment . . . And he had a marvellous sense of humour.[11]

It is difficult to imagine any other controller in the entire history of the KGB as ideally suited as Deutsch to the Cambridge Five. Though all but Philby graduated from Cambridge with first-class honours,[12] Deutsch's student career was even more brilliant than theirs, his understanding of human character much deeper, and his experience of life much broader. He combined a charismatic personality with visionary faith in the future of a human race freed from the exploitation and alienation of the capitalist system.

In July 1928, two months after his twenty-fourth birthday and less than five years after entering Vienna University as a first-year undergraduate, Deutsch was awarded the degree of PhD with distinction. Though his thesis had been on chemistry, Deutsch had also become deeply immersed in philosophy and psychology. His description of himself in university documents throughout his student years as a practising Jew (*mosaisch*)[13] was probably intended to conceal the fact that his religious faith had been replaced by an ardent commitment to the Communist International. Deutsch's vision of a new world order embraced sexual as well as political liberation. At about the time he began covert work for Comintern, he became publicly involved in the 'sex-pol' (sexual politics) movement, founded by the Viennese psychologist and sexologist Wilhelm Reich, which opened clinics to bring sexual enlightenment to Viennese workers.[14] At this stage of his career, Reich, then a Communist, was engaged in an ambitious attempt to integrate Freudianism with Marxism and in the early stages of a research programme on human sexual behaviour which later earned him a probably undeserved reputation as 'the prophet of the better orgasm'.[15] Deutsch enthusiastically embraced Reich's teaching that political and sexual repression were different sides of the same coin and together paved the way for fascism. He ran the Munster Verlag in Vienna which published Reich's work and other 'sex-pol' literature.[16] Though the Viennese police seems to have been unaware of his secret work for Soviet intelligence, its 'anti-pornography' section took an active interest in his involvement with the 'sex-pol' movement.[17]

Deutsch travelled to London in April 1934 under his real name, giving his profession as 'university lecturer' and using his academic credentials to mix in university circles. After living in temporary accommodation, he moved to a flat in the heartland of London's radical intelligentsia in Lawn Road, Hampstead. The 'Lawn Road Flats', as they were then known, were the first 'deck-access' apartments built in England (a type of construction later imitated in countless blocks of council flats) and, at the time, was probably Hampstead's most avant-garde building. Deutsch moved into number 7, next to the celebrated crime novelist Agatha Christie. While most of the residents' front doors were visible from the street, as is normal with the deck-access design, Deutsch's was concealed by a stairwell to make it possible for him and his visitors to enter and leave unobserved.[18]

Deutsch's message of liberation had all the greater appeal to the Cambridge Five because it had a sexual as well as a political dimension. All Five were rebels against the strict sexual mores as well as the antiquated class system of interwar Britain. Burgess and Blunt were homosexuals, Maclean a bisexual, and Philby a heterosexual athlete. Cairncross, a committed if unconventional heterosexual, later wrote a history of polygamy which concluded chauvinistically with a quotation from George Bernard Shaw: 'Women will always prefer a 10 per cent share of a first-rate man to sole ownership of a mediocre man.'[19] Cairncross plainly considered himself a first-rate rather than a mediocre man. Graham Greene was charmed by Cairncross's book. 'Here at last', he wrote to Cairncross, 'is a book which will appeal strongly to all polygamists.'[20]

Philby's first major service to Soviet intelligence was to direct Deutsch to other potential Cambridge recruits, chief among them Donald Maclean and Guy Burgess. If not already a committed Communist by the time he entered Trinity Hall, Cambridge, in 1931, Donald Maclean became one during his first year. As the handsome, academically gifted son of a former Liberal cabinet minister, Maclean must have seemed to Deutsch an almost ideal candidate to penetrate the corridors of power. On his graduation with first-class honours in modern languages in June 1934, however, Maclean showed no immediate sign of wanting a career in Whitehall. His ambition was either to teach English in the Soviet Union or to stay at Cambridge to work for a PhD. After being approached by Philby on Deutsch's instructions in August, he changed his mind, agreed to 'penetrate the bourgeois institutions' and began preparing for the Foreign Office entrance examinations in the following year.[21] Maclean's first codename, like Philby's, had two versions: WAISE in German, SIROTA in Russian – both meaning ORPHAN (an allusion to the death of his father in 1932).[22]

For some months Guy Burgess, then in his second year as a history research student at Trinity College preparing a thesis which he was never to complete, had been enthused by the idea of conducting an underground war against fascism on behalf of the Communist International. Ironically, in view of the fact that he was soon to become the third member of the 'Magnificent Five', he seems to have been inspired by the example of the *Fünfergruppen*, the secret 'groups of five' being formed by German Communists to organise opposition to Hitler. Maclean was, very probably, among the Communist friends with whom he discussed the (in reality rather unsuccessful) German groups.[23] When Maclean admitted, against his instructions, that he had been asked to engage in secret work,[24] Burgess was desperate for an invitation to join him.

In December 1934 Maclean arranged a first meeting between Deutsch and Burgess.[25] Deutsch already knew that Burgess was one of the most flamboyant figures in Cambridge: a brilliant, gregarious conversationalist equally at home in the teetotal intellectual discussions of the Apostles, the socially exclusive and heavy-drinking Pitt Club, and the irreverent satirical revues of the Footlights. He made no secret either of his Communist sympathies or of his enjoyment of the then illegal pleasures of homosexual 'rough trade' with young working-class men. A more doctrinaire and less imaginative controller than Deutsch might have concluded that the outrageous Burgess would be a liability rather than an asset. But Deutsch may well have sensed that Burgess's very outrageousness would give him good, if unconventional, cover for his work as a secret agent. No existing stereotype of a Soviet spy remotely resembled Burgess.[26] When invited to join the Comintern's underground struggle against fascism, Burgess told Deutsch that he 'was honoured and ready to sacrifice everything for the cause.' His codename MÄDCHEN (LITTLE GIRL, by contrast with Philby's codename SONNY) was a transparent reference to his homosexuality.[27]

In August 1935 Maclean passed the Foreign Office exams with flying colours. When asked about his 'Communist views' at Cambridge, Maclean decided to 'brazen it out':

'Yes,' I said, 'I did have such views – and I haven't entirely shaken them off.' I think they must have liked my honesty because they nodded, looked at each other and smiled. Then the chairman said: 'Thank-you, that will be all, Mr Maclean.'

In October 1935, as a new member of His Majesty's Diplomatic Service, Maclean became the first of the Magnificent Five to penetrate the corridors of power.[28]

Burgess went about burying his Communist past with characteristic flamboyance. Late in 1935 he became personal assistant to the young, right-wing gay Conservative MP, Captain 'Jack' Macnamara. Together they went on fact-finding missions to Nazi Germany which, according to Burgess, consisted largely of homosexual escapades with likeminded members of the Hitler Youth. Burgess built up a remarkable range of contacts among the continental 'Homintern'. Chief among them was Edouard Pfeiffer, chef de cabinet to Edouard Daladier, French War Minister from January 1936 to May 1940 and Prime Minister from April 1938 to March 1940. Burgess boasted to friends that, 'He and Pfeiffer and two members of the French Cabinet . . . had spent an evening together at a male brothel in Paris. Singing and dancing, they had danced around a table, lashing a naked boy, who was strapped to it, with leather whips.'[29]

During almost four years controlling British agents, Deutsch served under three illegal residents (Soviet heads of station without 'legal' diplomatic or other official cover), each of whom operated under a variety of aliases: Ignati Reif, codenamed MARR; Aleksandr Orlov, codenamed SCHWED (SWEDE); and Teodor Maly, codenamed PAUL, THEO, and MANN.[30] Reif had only a walk-on part in the history of the 'Magnificent Five'. Though Orlov spent only just over a year in Britain in 1934–5, he played a significant role in overseeing the recruitment of Philby, Maclean, and Burgess.[31] Only Maly, however, rivalled Deutsch's influence on the Cambridge recruits. Hungarian by birth, Maly entered a Catholic monastic order before the First World War but had volunteered for Military service in 1914.[32] He later told one of his agents:

> I saw all the horrors, young men with frozen limbs dying in the trenches . . . I lost my faith in God and when the Revolution broke out I joined the Bolsheviks. I broke with my past completely . . . I became a Communist and have always remained one.[33]

From April 1936 Maly shared in the running of the Cambridge agents, impressing them, like Deutsch, with both his human sympathy and his visionary faith in the Communist millennium.[34]

During the early months of 1937 Deutsch and Maly completed the recruitment of the 'Magnificent Five'. At the beginning of the year, Burgess, by then a producer at the BBC, arranged a first meeting between Deutsch and Anthony Blunt (codenamed TONY), French linguist, art historian and Fellow of Trinity College, Cambridge.[35] Though the title of 'Fourth Man' later accorded Blunt was a media invention rather than a KGB soubriquet, he was both the fourth of the 'Five' to be recruited and, over forty years later, the fourth to be publicly exposed. Until the war Blunt's chief role for the NKVD was that of talent-spotter.

The most talented potential agent identified by Blunt was the 'Fifth Man': John Cairncross, a brilliant Scot who had entered Trinity in 1934 at the age of twenty-one with a scholarship in modern languages, having already studied for two years at Glasgow University and gained a *license-ès-lettres* at the Sorbonne.[36] His passionate Marxism led the *Trinity Magazine* to give him the nickname 'The Fiery Cross', while his remarkable talent as a linguist led the same magazine to complain, 'Cairncross . . . learns a new language every fortnight.'[37] Among his College supervisors in French literature was Blunt, whom Cairncross found 'fascinating, charming and utterly ruthless'.[38] After graduating with first-class honours in 1936, Cairncross passed top of the Foreign Office entrance examinations, one hundred marks ahead of the next candidate (though he did less well at the interview).[39]

The initial approach to Cairncross early in 1937 was entrusted

by Deutsch to Burgess, much as Philby had made the first recruitment overture to Maclean in 1934. On 9 April 1937 Maly
informed Moscow that he had been formally recruited as a Soviet
agent codenamed MOLIERE.[40] (Cairncross was later to publish two
scholarly studies of Molière in the French language.) The last of
the Magnificent Five to be recruited, Cairncross was also the last
to be publicly exposed.[41]

All Five went on to achieve remarkable success in penetrating
Whitehall and the British intelligence services. Maclean continued to serve as a British diplomat until his defection to Moscow
in 1951. Cairncross moved from the Foreign Office to the
Treasury in 1938, then served successively as private secretary to
one of Churchill's ministers, Lord Hankey, with access to many
secret cabinet papers (1940–2), in Bletchley Park (1942–3), in the
Secret Intelligence Service, better known as SIS or MI6 (1943–5),
and again in the Treasury (1945–51). Burgess worked in SIS
(1938–40), the BBC (1940–4), and the Foreign Office (1944–51).
After a brief period in the Special Operations Executive in 1940,
Philby rose steadily up the ranks of SIS over the next decade,
being tipped by some as its future chief. Blunt worked in MI5 for
most of the Second World War and ran his former pupil Leo
Long as a sub-agent in military intelligence.

Many of the Five's successes on the eve of, and during, the
Second World War, however, were achieved despite, rather than
because of, their KGB controllers. By the end of 1937 the
inspirational leadership of Deutsch and Maly had been been
abruptly ended by Stalin's Great Terror. The paranoid hunt for
mostly imaginary traitors brought a large part of the Soviet
foreign intelligence network close to collapse. All three of the
illegal residents under whom Deutsch served in London were
'unmasked' as traitors. Reif and Maly were shot for imaginary
crimes. Orlov defected just in time to North America, securing
his survival by threatening to arrange for the revelation of all he
knew about Soviet espionage should he be pursued by an

NKVD assassination squad. Deutsch was recalled to Moscow, but was one of the few leading foreign intelligence officers to survive the Terror.[42] The Five lacked a regular controller until Anatoli Veniaminovich Gorsky was appointed London 'legal' resident under diplomatic cover late in 1940.[43]

The paranoia of the Great Terror degraded intelligence analysis as well as the running of the foreign agent network. Even after the Terror had abated, Stalin who frequently acted as his own chief intelligence analyst, continued to distrust intelligence from English sources. By the time of the conclusion of the Nazi–Soviet Pact in August 1939 he distrusted Neville Chamberlain and Winston Churchill more than Hitler. Stalin interpreted many of the more than a hundred warnings of Hitler's preparations for Operation BARBAROSSA, the invasion of the Soviet Union begun on 22 June 1941, as part of a Machiavellian plot by Churchill to engineer a Russo-German conflict. He scrawled in red ink on one such warning from Prague in April 1941: 'English provocation. Investigate! Stalin.'[44]

The Great Terror, the recall of Deutsch and Maly, and the Nazi–Soviet Pact degraded the Five as well as the KGB. All resorted to various forms of psychological denial in order to cling on to threadbare versions of the revolutionary dream which had turned them into Soviet agents. Philby claimed later that the 'only course of action open to me was to stick it out, in the confident faith that the principles of the revolution would outlive the aberration of individuals, however enormous.'[45] Cairncross subsequently became so ashamed of his continued work as a Soviet agent after the conclusion of the Nazi–Soviet pact that he insisted in his memoirs that he gave the KGB no further documents until the German invasion of Russia.[46] In reality he provided so many top-secret documents that the London residency complained they were too numerous to forward to Moscow by cipher telegram.[47] For all the Five, Hitler's surprise attack on the Soviet Union on 22 June 1941 came as an immense relief. Convinced that the Second World War would henceforth be decided on the Eastern Front, they convinced themselves anew of the rightness of their cause.

Among the most important top-secret documents available to Cairncross as Hankey's private secretary were the papers of the Scientific Advisory Committee (SAC), composed of some of Britain's most distinguished scientists, which met for the first time in October 1940 to coordinate the application of science to the war effort. Cairncross's access to SAC papers enabled him to give the KGB in September 1941 its first warning of plans to build the atomic bomb. In later years Cairncross was unable to come to terms with the fact that he had been the first of the atom spies, and denied in his memoirs that he had given any atomic intelligence to the Soviet Union. KGB archives prove that he did.[48]

For most of the Second World War, however, the Centre's ability to exploit the remarkable talents of the Five was seriously impeded by its paranoid tendencies. The intelligence they provided on British policy to the Soviet Union was simply not sinister enough for either Stalin or the Centre to find it credible. Early in April 1942 the Centre completed a lengthy analysis of the SIS documents and other highly classified intelligence supplied by Philby up to the end of 1941. Though praising SÖHNCHEN for 'systematically sending a lot of interesting material', it was puzzled that this material appeared to show that SIS had no agent network in Russia and was conducting only 'extremely insignificant' operations against the Soviet Union. KGB analysts had two reasons for disputing these entirely accurate conclusions. First, they were convinced that SIS had been conducting major operations inside the Soviet Union, using 'their most highly-skilled agents', throughout the 1930s. The reality – that SIS had not even possessed a pre-war Moscow station – was, so far as the Centre was concerned, literally unbelievable. Secondly, the Centre refused to believe that the Soviet Union was a smaller wartime priority for British intelligence (which was, in reality, almost wholly geared to the war effort) than Britain was for the KGB:

If the HOTEL [SIS] has recruited a hundred agents in Europe over the past few years, mainly from countries occupied by the Germans, there can be no doubt that our country receives no less attention.[49]

Like Philby, Maclean, Cairncross, Blunt and Burgess provided a remarkable amount of wartime intelligence. The problem for the professionally suspicious minds in the Centre was that it all seemed too good to be true. Taking their cue from the master conspiracy theorist in the Kremlin, they eventually concluded that what appeared to be the best intelligence ever obtained from Britain by any intelligence service was at root a British plot. The Five, later acknowledged as the ablest group of agents in KGB history, were discredited in the eyes of the wartime Centre leadership by their failure to provide evidence of a massive, non-existent British conspiracy against the Soviet Union. Of the reality of that conspiracy, even after the formation of the Grand Alliance, Stalin and his chief intelligence advisers had no doubt. In October 1942 Stalin wrote to the Soviet ambassador in Britain, Ivan Maisky:

> All of us in Moscow have gained the impression that Churchill is aiming at the defeat of the USSR, in order then to come to terms with Germany . . . at the expense of our country.[50]

On 25 October 1943 the Centre informed the London residency that it was now clear, after long analysis of the voluminous intelligence from the Five, that they were double agents, working on the instructions of SIS and MI5. As far back as their years at Cambridge, Philby, Maclean and Burgess had probably been acting on instructions from British Intelligence to infiltrate the student Left before making contact with the NKVD. Only thus, the Centre reasoned, was it possible to explain why both SIS and MI5 were currently employing in highly sensitive jobs Cambridge graduates with a Communist background. The content of the intelligence supplied by SÖHNCHEN (Philby) from SIS and by TONY (Blunt) from MI5 was further evidence that they were being used to feed disinformation to the NKGB:

During the entire period that s[öhnchen] and t[ony] worked for the British special services, they did not help expose a single valuable ISLANDERS [British] agent either in the USSR or in the Soviet embassy in the ISLAND [Britain].

There was, of course, no such 'valuable agent' for Philby in SIS or Blunt in MI5 to expose, but that simple possibility did not even occur to the hardened conspiracy theorists in the Centre. Philby's accurate report that 'at the present time the HOTEL [SIS] is not engaged in active work against the Soviet Union' was also, in the Centre's view, obvious disinformation.[51]

Even the Centre, however, had some difficulty in explaining why the Five were providing, along with disinformation, large amounts of apparently accurate high-grade intelligence. In its missive to the London residency of 25 October, the Centre suggested a number of possible answers to this baffling problem. The sheer quantity of Foreign Office documents supplied by Maclean *might* indicate, it believed, that, unlike the other four, he was not *consciously* deceiving the NKGB, but was merely being manipulated by the others to the best of their ability. The Centre also argued that the Five were instructed to pass on important intelligence about Germany which did not harm British interests in order to make their disinformation about British policy more credible.[52]

The most valuable intelligence on Germany to reach the Centre from London in 1943 were German decrypts supplied by Cairncross from Bletchley Park. A brief official biography of the wartime foreign intelligence chief, Pavel Mikhailovich Fitin, published by the SVR in 1995, singles out for special mention the ULTRA intelligence obtained from Britain on German preparations for the battle of Kursk when the Red Army halted Operation CITADEL, Hitler's last major offensive on the Eastern Front.[53] The *Luftwaffe* decrypts provided by Cairncross were of crucial importance in enabling the Red Air Force to launch massive preemptive strikes against German airfields which destroyed over 500 enemy aircraft.[54] The Centre's addiction to conspiracy theory ran so deep, however, that it was capable of regarding the agent who supplied intelligence of critical importance before Kursk as part of an elaborate network of deception.

Not till after D-Day and the opening of the Second Front did the Centre's suspicions of British policy decline to a level which made it possible for them to grasp that the Five were, in reality, working for them. On 29 June 1944 the Centre informed the London residency, now headed by Konstantin Mikhailovich Kukin, that recent important SIS documents provided by Philby had been largely corroborated by material from 'other sources' (some probably in the American OSS, with whom SIS exchanged many highly classified reports[55]): 'This is a serious confirmation of s[ÖHNCHEN]'s honesty in his work with us, which obliges us to review our attitude towards him and the entire group.' It was now clear, the Centre acknowledged, that intelligence from the Five was 'of great value', and contact with them must be maintained at all costs:

> On our behalf express much gratitude to S[ÖHNCHEN] for his work
> ... If you find it convenient and possible, offer S[ÖHNCHEN] in the
> most tactful way a bonus of £100 or give him a gift of equal value.

After six years in which his phenomenal work as a penetration agent had been frequently undervalued, ignored or suspected by the Centre, Philby was almost pathetically grateful for the long overdue recognition of his achievements. 'During this decade of work', he told Moscow, 'I have never been so deeply touched as now with your gift and no less deeply excited by your communication [of thanks].'[56]

At about the same time that Philby was given his present, Cairncross was belatedly rewarded for his contribution to the epic victory at Kursk. His new controller, Boris Krötenschield, informed him that he had been awarded one of the highest Soviet decorations, the Order of the Red Banner. He opened a velvet-lined box, took out the decoration, and placed it Cairncross's hands. Krötenschield reported to the Centre that Cairncross was visibly elated by the award, though he was told to hand it back for safe-keeping in Moscow.[57]

The extraordinary first decade of the Magnificent Five illustrates the gulf which, particularly in authoritarian regimes, sometimes separates intelligence collection from intelligence analysis. Thanks chiefly to the Five, the KGB received better intelligence from Britain than any power had ever received before (or, probably, since). Because of the paranoid tendencies inherent in the Stalinist regime's view of the outside world, however, it frequently failed to comprehend its significance. Thus, incredibly, in the middle of the Second World War, the KGB mistook probably the best foreign agents in its history for a deep-laid deception plot by British intelligence.

NOTES

1 Sir F. H. Hinsley and Alan Stripp (eds.), *Codebreakers: The Inside Story of Bletchley Park* (Oxford: Oxford University Press, 1993).

2 Before 1954 the KGB was known successively as the Cheka, the GPU, the OGPU, the GUGB (of the NKVD), the NKGB and the MGB. To avoid confusion it is referred to throughout this article as the KGB. For details of its nomenclature, see Christopher Andrew and Oleg Gordievsky, *KGB: The Inside Story of Its Foreign Operations from Lenin to Gorbachev*, paperback edition (London: Sceptre, 1991), p. 14.

3 Studies of the Magnificent Five during the 1990s have, for the first time, been able to draw, in varying degrees, on KGB archives. Andrew and Gordievsky, *KGB* (first published in 1990) drew on Gordievsky's access to the operational files of Blunt and his sub-agent, Leo Long, and to other classified records relating to the Five in the KGB First Chief Directorate Memory Room and elsewhere. The files smuggled out of the KGB by Gordievsky, of which two volumes have so far been published, did not, however, include any on the Five. The authors of three volumes authorised by the SVR (today's

Russian foreign intelligence service) were given privileged, partial access to relevant KGB records: John Costello and Oleg Tsarev, *Deadly Illusions* (London: Century, 1993); Genrikh Borovik, *The Philby Files* (London: Little, Brown, 1994); Nigel West and Oleg Tsarev, *The Crown Jewels* (London: Harper Collins, 1998).

4 Andrew and Gordievsky, *KGB*, pp. 198–9, 209–10.

5 Paul Hollander, *Political Pilgrims* (Oxford: Oxford University Press, 1981), p. 102.

6 Andrew and Gordievsky, *KGB*, pp. 209–13. Costello and Tsarev, *Deadly Illusions*, pp. 125–30.

7 Andrew and Gordievsky, *KGB*, pp. 210–11. Phillip Knightley, *Philby: KGB Masterspy* (London: André Deutsch, 1988), ch. 3.

8 Costello and Tsarev, *Deadly Illusions*, pp. 132–4.

9 Ibid. pp. 134–6.

10 The text of the report on Deutsch's first meeting with Philby, sent to the Centre by the London illegal resident, Ignati Reif, is published in Borovik, *Philby Files*, pp. 38–40. Cf. Costello and Tsarev, *Deadly Illusions*, p. 137.

11 Borovik, *Philby Files*, p. 29.

12 Burgess gained a first in Part I History but was ill during Part II and awarded an aegrotat (the unclassed honours awarded to those unable to sit their examinations for medical reasons).

13 'Nationale für ordentliche Hörer der philosophischen Fakultät': entries for Arnold Deutsch, 1923–7; 'Rigorosenakt des Arnold Deutsch', 1928, no. 9929, with c.v. by Deutsch; records of Deutsch's 1928 PhD examination. Archives of University of Vienna.

14 Andrew and Gordievsky, *KGB*, pp. 214–15.

15 Myron Sharaf, *Fury on Earth: A Biography of Wilhelm Reich* (London: André Deutsch, 1983).

16 Wilhelm Reich, *Sexualerregung und Sexualbefriedigung*, the first publication in the series *Schriften der Sozialistischen Gesellschaft für Sexualberatung und Sexualforschung in Wien*, carries the note 'Copyright 1929 by Münster-Verlag (Dr Arnold Deutsch), Wien II'.

17 Viennese police reports on Deutsch of 25 March and 27 April 1934 (ref. Z1.38.Z.g.p./34), Dokumentationsarchiv des Österreichischen Widerstandes, Vienna.

18 Deutsch's address and profession as 'university lecturer' are given on

the birth-certificate of his daughter, Ninette Elizabeth, born on 21 May 1936. Further information from residents of Lawn Road Flats.

19 John Cairncross, *After Polygamy Was Made a Sin* (London: Routledge, 1974).

20 Cairncross quotes Greene's letter to him in a postscript to his book *La Fontaine Fables and Other Poems* (Gerrards Cross: Colin Smythe, 1982).

21 Andrew and Gordievsky, *KGB*, pp. 223–6.

22 Costello and Tsarev, *Deadly Illusions*, pp. 186–8.

23 Andrew and Gordievsky, *KGB*, pp. 206–8.

24 Costello and Tsarev, *Deadly Illusions*, p. 224.

25 Ibid., *Deadly Illusions*, p. 225.

26 Andrew and Gordievsky, *KGB*, pp. 216–19.

27 Costello and Tsarev, *Deadly Illusions*, ch. 9.

28 Andrew and Gordievsky, *KGB*, pp. 225–6.

29 Goronwy Rees, *A Chapter of Accidents* (London: Chatto & Windus, 1971), pp. 122–3. Michael Straight, *After Long Silence* (London: Collins, 1983), pp. 94–5, 142.

30 Both their photographs and codenames appear in Costello and Tsarev, *Deadly Illusions*.

31 Somewhat misleadingly, Costello and Tsarev, *Deadly Illusions* (a KGB/SVR-sponsored biography of Orlov) claimed that Orlov was 'the mastermind' responsible for the recruitment of the Cambridge agents. One reason for this exaggeration is hierarchical. Within the Soviet *nomenklatura* senior bureaucrats commonly claimed, and were accorded, the credit for their subordinates' successes. The recruitment of Philby is a characteristic example of this common phenomenon. In a number of respects the detailed evidence advanced by Costello and Tsarev is at odds with its overstatement of Orlov's importance by comparison with Deutsch. At one point (p. 193) the authors themselves acknowledge: 'A search through the archives of the seventy-five year history of the Soviet intelligence service reveals that few officers came close to matching Deutsch's record.' Orlov was not one of the few who did.

32 Though some of his agents believed that Maly had been a Catholic priest, his operational file shows that he had only deacon's orders when he volunteered for the army. West and Tsarev, *Crown Jewels*, pp. 113–14.

33 Elizabeth Poretsky, *Our Own People* (London: Oxford University Press, 1969), pp. 214–15. Igor Cornelissen, *De GPOe op de Overtoom* (Amsterdam: Van Gennep, 1989), ch. 11.

34 Andrew and Gordievsky, *KGB*, pp. 211–13, 229–30. Costello and Tsarev, *Deadly Illusions*, pp. 199ff.

35 Costello and Tsarev, *Deadly Illusions*, p. 245.

36 Details of Cairncross's academic career from the archives of Glasgow University, Trinity College, Cambridge, and Cambridge University.

37 *Trinity Magazine*, Easter Term 1935 and Easter Term 1936.

38 Barrie Penrose and Simon Freeman, *Conspiracy of Silence* (London: Grafton Books, 1986), pp. 369–71.

39 Sir John Colville, *The Fringes of Power* (London: Hodder & Stoughton, 1985), p. 30n. John Cairncross's unreliable memoirs (*The Enigma Spy* [London: Century, 1997]) are almost a textbook case of psychological denial. At almost every stage of his career as a Soviet agent (save for a heroic year at Bletchley Park in 1942–3, when he claims to have been instrumental in 'changing the course of World War Two'), Cairncross seeks to diminish or deny the significance of his role. His version of his career as a Soviet agent, save for the year at Bletchley Park, is comprehensively contradicted by the evidence of the KGB files.

40 The former Chairman of the Cambridge University Communist Party, James Klugmann, completed the recruitment of Cairncross begun by Burgess. Costello and Tsarev, *Deadly Illusions*, p. 214; West and Tsarev, *Crown Jewels*, pp. 204–8.

41 Though publicly identified as a Soviet agent in 1980, Cairncross's role as the Fifth Man was not revealed until the publication in 1990 of Andrew and Gordievsky, *KGB*. During the 1980s the media hunt for the Fifth Man and other Soviet agents sometimes resembled Monty Python's quest for the Holy Grail. Among those mistakenly accused were Frank Birch, Sefton Delmer, Andrew Gow, Sir Roger Hollis, Guy Liddell, Graham Mitchell, and Arthur Pigou, all deceased; Sir Rudolf Peierls, who denied claims that he too was dead and sued successfully for libel; Lord Rothschild, the victim until his death of innuendo rather than open accusation in case he also sued; and Dr Wilfred Mann, who did not sue but published a book proving his innocence.

42 Andrew and Gordievsky, *KGB*, pp. 234–6. Costello and Tsarev, *Deadly Illusions*. Deutsch died in 1942 on his way to a posting in the United States when his ship was sunk by a German U-boat. T. V. Samolis (ed.), *Veterany Vneshnei Razvedki Rossii: Kratkiy Biografichesky Spravochnik* (Moscow: SVR Press, 1995), p. 42.

43 Gorsky was one of the seventy-five heroes of foreign intelligence selected by the SVR for a volume of brief hagiographies to commemorate the seventy-fifth anniversary of the founding of Soviet foreign intelligence; Samolis, *Veterany Vneshnei Razvedki Rossii*, pp. 31–2.

44 Andrew and Gordievsky, *KGB*, ch. 7.

45 Kim Philby, *My Silent War*, paperback edition (London: Granada, 1969), p. 16.

46 Cairncross, *The Enigma Spy*, p. 82.

47 Andrew and Gordievsky, *KGB*, p. 272. West and Tsarev, *Crown Jewels*, pp. 214–15.

48 The revelation that Cairncross, thanks to his access to Scientific Advisory Committee papers, was the first to warn the Centre of the plan to construct the atomic bomb first appeared in 1990 in Andrew and Gordievsky, *KGB*, p. 321. For further revelations from KGB archives on his role as the first of the atom spies see: West and Tsarev, *Crown Jewels*, pp. 228, 234; Michael Smith, 'The humble Scot who rose to the top – but then chose treachery', *Daily Telegraph*, 12 January 1998.

49 Borovik, *Philby Files*, pp. 196–7. On SIS's lack of a Moscow station in the the 1930s, see Christopher Andrew, *Secret Service: The Making of the British Intelligence Community*, paperback edition (London: Sceptre, 1986), p. 573.

50 Jonathan Haslam, 'Stalin's Fears of a Separate Peace 1942', *Intelligence and National Security*, 8 (1993), pp. 97–9.

51 Borovik, *Philby Files*, pp. 216–18.

52 Ibid., p. 217n.

53 Samolis, *Veterany Vneshnei Razvedki Rossii*, p. 154.

54 Cairncross, *The Enigma Spy*, ch. 7; Andrew and Gordievsky, *KGB*, p. 314; West and Tsarev, *Crown Jewels*, p. 220.

55 Among the documents Philby passed to the KGB were the German Foreign Ministry documents obtained by OSS in Switzerland and probably also supplied by NKGB agents in OSS. Philby, *My Silent*

War, pp. 84–6. Christopher Andrew, *For The President's Eyes Only: Secret Intelligence and the American Presidency from Washington to Bush* (London: Harper Collins, 1995), pp. 141–2.

56 Borovik, *Philby Files*, pp. 232–3.

57 Yuri Modin, *My Five Cambridge Friends* (London: Headline, 1994), p. 114. From 1944 to 1947 Modin was responsible for the files of the Five at the Centre, before being posted to London to act as their controller.

Printed in the United States
By Bookmasters